Goodwill's Skill Builders

1000 MATH PROBLEMS

GOODWILL'S
SKILL BUILDERS
1000
MATH PROBLEMS

FRACTIONS ☆ DECIMALS ☆ PERCENTS
☆ WORD PROBLEMS ☆ RATIOS
☆ PROPORTIONS ☆ PERIMETERS
☆ AREAS ☆ VARIABLES
☆ DISTANCE, RATE AND TIME

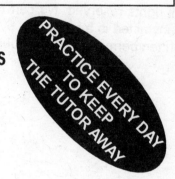

SAM PHILIPS

GOODWILL PUBLISHING HOUSE®
B-3 RATTAN JYOTI, 18 RAJENDRA PLACE
NEW DELHI -110008 (INDIA)

© Publishers

All rights reserved. No part of this publication may be reproduced, stored in a retrieval system transmitted in any form or by any means, mechanical, photocopying or otherwise, without the p written permission of the publisher.

Published by
GOODWILL PUBLISHING HOUSE®
B-3 Rattan Jyoti, 18 Rajendra Place
New Delhi-110008 (INDIA)
Tel : 25750801, 25820556
Fax : 91-11-25764396
E-mail : goodwillpub@vsnl.net
Website : www.goodwillpublishinghouse.com

Printed at: Batra Art Press, New Delhi-28

INTRODUCTION

Ask a question. Is Mathematics a tough subject ? Many students would reply affirmatively. But the reality is not that. It requires more practice. That is the reason why some say mathematics is the easiest subject. Most of the time there are not enough examples, exercises and efforts to improve one's math skills. The tendency is to solve only model questions. Remember, practice makes perfect !

This book contains 1000 objective type questions. Various topics like whole numbers, decimals, fractions, algebra, geometry, etc. are included in this book. The most important feature of this book is the fact that it is designed keeping in mind the students having different IQ. It generates curiosity and interest. With this book, problem solving becomes a habit and a matter of interest.

Unlike some guides, this book gives answers with an explanation. The healthy way of learning is to solve a set of questions and refer the answers given at the back. Whether learning or just brushing up your skills. It will be of immense help to train others. Math teachers will also find this book very useful. In fact, this makes their job simple.

The basic purpose of this book will be defeated if the students use calculators for solving their problems. Therefore, avoid calculators to improve skills.

CONTENTS

SECTION 1 : Miscellaneous Math 1

SECTION 2 : Fractions 25

SECTION 3 : Decimals 53

SECTION 4 : Percentages 75

SECTION 5 : Basic Algebra 99

SECTION 6 : Basic Geometry 123

ANSWERS 157

SECTION

1

MISCELLANEOUS MATH

The following section consists of 10 sets of miscellaneous math, including basic arithmetic questions and word problems with whole numbers. You'll also see problems involving pre-algebra concepts such as negative numbers, exponents, and square roots (getting you ready for the algebra in Section 5). This section will provide a warm-up session before you move on to more difficult kinds of problems.

1000 MATH PROBLEMS >>> *Miscellaneous Math*

SET 1

1. 7 + 3 =
   - (a) 4
   - (b) 8
   - (c) 9
   - (d) 10

2. 9 − 2 =
   - (a) 7
   - (b) 8
   - (c) 6
   - (d) 5

3. 8 ÷ 2 =
   - (a) 6
   - (b) 4
   - (c) 5
   - (d) 7

4. 3 × 3 =
   - (a) 6
   - (b) 7
   - (c) 8
   - (d) 9

5. 54 − 17 =
   - (a) 37
   - (b) 39
   - (c) 41
   - (d) 43

6. 62 + 7 =
   - (a) 64
   - (b) 68
   - (c) 69
   - (d) 70

7. 25 + 16 =
   - (a) 37
   - (b) 41
   - (c) 45
   - (d) 51

8. 23 + 22 =
   - (a) 45
   - (b) 55
   - (c) 61
   - (d) 62

9. 39 − 18 =
   - (a) 7
   - (b) 11
   - (c) 16
   - (d) 21

10. 36,785 − 188 =
    - (a) 35,697
    - (b) 36,497
    - (c) 36,597
    - (d) 37,007

11. 72 + 98 − 17 =
    - (a) 143
    - (b) 153
    - (c) 163
    - (d) 170

12. 376 − 360 + 337 =
    - (a) 663
    - (b) 553
    - (c) 453
    - (d) 353

13. 444 + 332 − 216 =
    - (a) 312
    - (b) 450
    - (c) 560
    - (d) 612

1000 MATH PROBLEMS >>> *Miscellaneous Math*

14. 2,710 + 4,370 − 400 =
   (a) 6,780
   (b) 6,680
   (c) 6,580
   (d) 6,480

15. 7,777 − 3,443 + 1,173 =
   (a) 5,507
   (b) 5,407
   (c) 5,307
   (d) 5,207

16. 5,776 − 2,227 − 1,993 =
   (a) 1,443
   (b) 1,476
   (c) 1,543
   (d) 1,556

1000 MATH PROBLEMS >>> *Miscellaneous Math*

SET 2

17. (25 + 17) (64 − 49) =
    (a) 57
    (b) 63
    (c) 570
    (d) 630

18. 42 ÷ 6 =
    (a) 8
    (b) 7
    (c) 6
    (d) 5

19. 292 × 50 =
    (a) 14,600
    (b) 14,500
    (c) 10,500
    (d) 1,450

20. 400 × 76 =
    (a) 52,000
    (b) 30,100
    (c) 30,400
    (d) 20,400

21. 2,273 × 4 =
    (a) 9,092
    (b) 8,982
    (c) 8,892
    (d) 8,882

22. 52,834 ÷ 9 is most nearly equal to
    (a) 5,870
    (b) 5,826
    (c) 5,826
    (d) 5,871

23. 70,014 ÷ 14 =
    (a) 5,001
    (b) 5,011
    (c) 5,111
    (d) 5,101

24. 703 × 365 =
    (a) 67,595
    (b) 255,695
    (c) 256,595
    (d) 263,595

25. 2,850 ÷ 190 =
    (a) 15
    (b) 16
    (c) 105
    (d) 150

26. 62,035 ÷ 5 =
    (a) 1,247
    (b) 12,470
    (c) 12,407
    (d) 13,610

27. 7,409 ÷ 74 =
    (a) 1
    (b) 10
    (c) 100
    (d) 1000

28. 4,563 × 45 =
    (a) 205,335
    (b) 206,665
    (c) 207,305
    (d) 206,335

29. 12(84 − 5) − (3 × 54) =
    (a) 54,000
    (b) 841
    (c) 796
    (d) 786

1000 MATH PROBLEMS >>> *Miscellaneous Math*

30. (2 + 4) × 8 =
    (a) 84
    (b) 64
    (c) 48
    (d) 32

31. (14 × 7) + 12 =
    (a) 98
    (b) 266
    (c) 110
    (d) 100

32. (667 × 2) + 133 =
    (a) 1,467
    (b) 1,307
    (c) 1,267
    (d) 1,117

1000 MATH PROBLEMS >>> Miscellaneous Math

SET 3

33. (403 × 163) − 678 =
    (a) 11,003
    (b) 34,772
    (c) 65,011
    (d) 67,020

34. (500 ÷ 20) + 76 =
    (a) 221
    (b) 201
    (c) 121
    (d) 101

35. 604 − 204 ÷ 2 =
    (a) 502
    (b) 301
    (c) 201
    (d) 101

36. 604 − (202 ÷ 2) =
    (a) 201
    (b) 302
    (c) 402
    (d) 503

37. 37 × 60 × 3 =
    (a) 3,660
    (b) 4,990
    (c) 5,760
    (d) 6,660

38. 12(9 × 4) =
    (a) 432
    (b) 336
    (c) 108
    (d) 72

39. Expanded form of 20,706
    (a) 200 + 70 + 6
    (b) 2,000 + 700 + 6
    (c) 20,000 + 70 + 6
    (d) 20,000 + 700 + 6

40. Which of the following choices is divisible by 6 and 7 ?
    (a) 63
    (b) 74
    (c) 84
    (d) 96

41. Which of the following means $5n + 7 = 17$?
    (a) 7 more than 5 times a number is 17
    (b) 5 more than 7 times a number is 17
    (c) 7 less than 5 times a number is 17
    (d) 12 times a number is 17

42. 184 is evenly divisible by
    (a) 46
    (b) 43
    (c) 41
    (d) 40

43. 4 pounds 14 ounces − 3 pounds 12 ounces =
    (a) 2 pounds 2 ounces
    (b) 1 pound 12 ounces
    (c) 1 pound 2 ounces
    (d) 14 ounces

44. 3 hours 20 minutes − 1 hour 48 minutes =
    (a) 5 hours 8 minutes
    (b) 4 hours 8 minutes
    (c) 2 hours 28 minutes
    (d) 1 hour 32 minutes

1000 MATH PROBLEMS >>> *Miscellaneous Math*

45. The average of 54, 61, 70 and 75 is
- (a) 55
- (b) 62
- (c) 65
- (d) 68

46. 52,830 ÷ 9 =
- (a) 5,870
- (b) 5,820
- (c) 5,826
- (d) 5,871

47. 2 feet 4 inches + 4 feet 8 inches =
- (a) 6 feet 8 inches
- (b) 7 feet
- (c) 7 feet 12 inches
- (d) 8 feet

48. 1 hour 20 minutes + 3 hours 30 minutes =
- (a) 4 hours
- (b) 4 hours 20 minutes
- (c) 4 hours 50 minutes
- (d) 5 hours

SET 4

49. What is the estimated product when 157 and 817 are rounded to the nearest hundred and multiplied?
    (a) 160,000
    (b) 180,000
    (c) 16,000
    (d) 80,000

50. $(-12) + 4 =$
    (a) zero
    (b) –4
    (c) –8
    (d) –16

51. $-4 \times (-6) =$
    (a) –10
    (b) 10
    (c) –24
    (d) 24

52. $-4 + (-4) =$
    (a) 16
    (b) –16
    (c) 8
    (d) –8

53. $6 + (-10)$
    (a) 4
    (b) –4
    (c) 16
    (d) –16

54. $16 + (-16) =$
    (a) zero
    (b) 32
    (c) –32
    (d) 16

55. What is another way to write $4 \times 4 \times 4$?
    (a) 3×4
    (b) 8×4
    (c) 4^3
    (d) 3^4

56. What is another way to write 3^4?
    (a) 12
    (b) 24
    (c) 27
    (d) 81

57. $6^3 =$
    (a) 36
    (b) 1,296
    (c) 18
    (d) 216

58. $17^2 =$
    (a) 34
    (b) 68
    (c) 136
    (d) 289

59. $10^5 \div 10^2 =$
    (a) 1^3
    (b) 10^3
    (c) 10^7
    (d) 10^{10}

60. $-6^2 =$
    (a) –36
    (b) 36
    (c) –12
    (d) 12

1000 MATH PROBLEMS >>> Miscellaneous Math

61. $-3^3 =$
    - (a) −9
    - (b) 9
    - (c) −27
    - (d) 27

62. $-12^2 =$
    - (a) 144
    - (b) −144
    - (c) −24
    - (d) 24

63. What is the square root of 16?
    - (a) 32
    - (b) 8
    - (c) 4
    - (d) 2

64. Which of these equation is INCORRECT?
    - (a) $\sqrt{16} + \sqrt{3} = \sqrt{16+3}$
    - (b) $\sqrt{6} \times \sqrt{12} = \sqrt{6 \times 12}$
    - (c) neither is incorrect
    - (d) both are incorrect

1000 MATH PROBLEMS ›»› *Miscellaneous Math*

SET 5

65. What is the square root of 64?
    (a) 16
    (b) 12
    (c) 8
    (d) 6

66. Hari got a box containing 6,000 kg silver from his ancestral house. The box was packed in 5 kg bags. How many bags of silver did the box contain?
    (a) 1,200 bags
    (b) 600 bags
    (c) 300 bags
    (d) 20 bags

67. Raju's horse, Spike, can run 3 times faster than Varun's horse, Muffin. The best simplification of this problem would be written
    (a) M ÷ 3 = 2
    (b) S = M − 2
    (c) M = S + 3
    (d) S = M × 3

68. Rita and Santosh bought 15 metres 30 centimeters and 20 metres 55 centimetres lace respectively. How much lace did they buy altogether?
    (a) 37
    (b) 35.85
    (c) 36.25
    (d) 35

69. Prakash rented two movies to watch last night. The first was 1 hour 40 minutes long, the second 1 hour 50 minutes long. How much time did it take for Prakash to watch the two videos?
    (a) 4.5 hours
    (b) 3.5 hours
    (c) 2.5 hours
    (d) 1.5 hours

70. The Neighbourhood Association organized a play group for cats. 15 people joined, and each brought 3 cats. How many cats were brought to the play group?
    (a) 60 cats
    (b) 45 cats
    (c) 30 cats
    (d) 25 cats

71. Raju had Rs 47 in his purse. He buys Apples for Rs. 15. Then he remembers his daughter asked for chocolate and he gets some for Rs. 5. How much money Raju Saves?
    (a) Rs. 30
    (b) Rs. 27
    (c) Rs. 25
    (d) Rs. 17

72. In a cotton mill on 8 hours work schedule, Seema spins 152 cotton reels, Salma spins 168 cotton reels and Savitri and Santosh spins 182 and 201 cotton reels respectively. How much reels together they produce that day?
    (a) 690 reels
    (b) 703 reels
    (c) 710 reels
    (d) 750 reels

1000 MATH PROBLEMS >>> Miscellaneous Math

73. Officer Tina has responded to the scene of a robbery. On the officer's arrival, the victim, Ms. Shama, tells the officer that the following items were taken from her by a man who threatened her with a knife:

    - 1 gold watch, valued at Rs 2400
    - 2 rings, each valued at Rs 1500
    - 1 ring, valued at Rs 700
    - Cash Rs 950

    Officer Tina is preparing her report on the robbery. Which one of the following is the total value of the cash and property Ms. Shama reported stolen?

    (a) Rs 5,450
    (b) Rs 5,550
    (c) Rs 7,050
    (d) Rs 7,850

74. The drivers at G & G trucking must report the mileage on their trucks each week. The mileage reading of Shahid's vehicle was 20,907 at the beginning of one week, and 21,053 at the end of the same week. What was the total number of miles driven by Shahid that week?

    (a) 46 miles
    (b) 145 miles
    (c) 146 miles
    (d) 1046 miles

75. In a downtown department store, Kavita finds a woman's handbag and turns it into the clerk in the Lost and Found Department. The clerk estimates that the handbag is worth approximately Rs 150. Inside, she finds the following items:

    - 1 leather makeup case valued at   Rs 65
    - 1 vial of perfume, unopened, valued at   Rs 75
    - 1 pair of earrings valued at   Rs 150
    - Cash   Rs 178

    The clerk is writing a report to be submitted along with the found property. What should she write as the total value of the found cash and property?

    (a) Rs 468
    (b) Rs 608
    (c) Rs 618
    (d) Rs 718

76. A trash container, when empty, weighs 27 kgs. If this container is filled with a load of trash that weighs 108 kgs, what is the total weight of the container and its contents?

    (a) 81 kgs
    (b) 135 kgs
    (c) 145 kgs
    (d) 185 kgs

1000 MATH PROBLEMS >>> Miscellaneous Math

77. Mrs. Das went to buy some kitchen wares. She purchased the following items:
    - 3 hot cases each costing Rs. 125
    - 2 big flasks each costing Rs. 1300
    - 2 Shelfs, each costing Rs. 950
    - One hand mixie Rs. 600
    - One steel box Rs. 50

    Write the total value of the items she purchased?
    - (a) Rs. 3,025
    - (b) Rs. 5,400
    - (c) Rs. 5,525
    - (d) Rs. 6,525

78. Mr Raju is purchasing gifts for his family. He stops to consider what else he has to buy. A quick mental inventory of his shopping bag so far reveals the following:
    - 1 Kashmeree sweater valued at Rs 260
    - 3 bracelets, each valued at Rs 365
    - 1 computer game valued at Rs 78
    - 1 brooch valued at Rs 130

    Later, having coffee in the Food Court, he suddenly remembers that he has purchased only 2 bracelets, not 3, and that the Kashmeree sweater was on sale for Rs 245. What is the total value of the gifts Mr. Raju has purchased so far?
    - (a) Rs 833
    - (b) Rs 1,183
    - (c) Rs 1,198
    - (d) Rs 1.563

79. Department regulations require trash collection trucks to have transmission maintenance every 13,000 miles. Truck #B-17 last had maintenance on its transmission at 12,398 miles. The mileage gauge now reads 22,003. How many more miles can the truck be driven before it must be brought in for transmission maintenance ?
    - (a) 3,395 miles
    - (b) 4,395 miles
    - (c) 9,003 miles
    - (d) 9,605 miles

80. The city's bus system carries 1,200,000 people each day. How many people does the bus system carry each year? (1 year = 365 days)
    - (a) 3,288 people
    - (b) 32,880 people
    - (c) 43,800,000 people
    - (d) 438,000,000 people

1000 MATH PROBLEMS >>> Miscellaneous Math

SET 6

81. How many inches are there in 4 feet?
    (a) 12 inches
    (b) 36 inches
    (c) 48 inches
    (d) 52 inches

82. Ravi is 46 years old, twice as old as Rajeeve. How old is Rajeeve?
    (a) 30 years old
    (b) 28 years old
    (c) 23 years old
    (d) 18 years old

83. Salesperson Rita drives 2,052 miles in 6 days, stopping at 2 towns each day. How many miles does she average between stops?
    (a) 171 miles
    (b) 342 miles
    (c) 684 miles
    (d) 1,026 miles

84. During the last week of track training, Shaloo achieves the following times in seconds: 66, 57, 54, 55, 64, 59 and 61. Her three best times this week are averaged for her final score on the course. What is her final score?
    (a) 57 seconds
    (b) 55 seconds
    (c) 59 seconds
    (d) 61 seconds

85. A train must travel to a certain town in six days. The town is 3,450 miles away. The train must travel an average of how many miles each day to reach its destination?
    (a) 500 miles
    (b) 525 miles
    (c) 550 miles
    (d) 575 miles

86. On the cardiac ward of a metropolitan hospital, there are 7 nursing assistants. NA Rama has 8 patients; NA Rani has 5 patients; NA Manju has 9 patients; NA Hira has 10 patients; NA Gauri has 10 patients; NA Jamuna has 14 patients, and NA Dinesh has 7 patients. What is the average number of patients per nursing assistant?
    (a) 7 patients
    (b) 8 patients
    (c) 9 patients
    (d) 10 patients

87. A car uses 16 gallons of gas to travel 448 miles. How many miles per gallon does the car get?
    (a) 22 miles per gallon
    (b) 24 miles per gallon
    (c) 26 miles per gallon
    (d) 28 miles per gallon

88. A floppy disk shows 827,036 bytes free and 629,352 bytes used. If you delete a file of size 542,159 bytes and create a new file of size 489,986 bytes, how many free bytes will the floppy disk have?
    (a) 577,179 free bytes
    (b) 681,525 free bytes
    (c) 774,863 free bytes
    (d) 879,209 free bytes

1000 MATH PROBLEMS >>> Miscellaneous Math

89. A family's gas and electricity bill averages Rs. 80 a month for seven months of the year and Rs. 20 a month for the rest of the year. If the family's bills were averaged over the entire year, what would the monthly bill be?
(a) Rs. 45
(b) Rs. 50
(c) Rs. 55
(d) Rs. 60

90. If a vehicle is driven 22 miles on Monday, 25 miles on Tuesday, and 19 miles on Wednesday, what is the average number of miles driven each day?
(a) 19 miles
(b) 21 miles
(c) 22 miles
(d) 23 miles

91. A piece of gauze 3 feet 4 inches long was divided in 5 equal parts. How long was each part?
(a) 1 foot 2 inches
(b) 10 inches
(c) 8 inches
(d) 6 inches

92. Each sprinkler head in a hospital ward sprinkler system sprays water at an average of 16 gallons per minute. If 5 sprinkler heads are flowing at the same time, how many gallons of water will be released in 10 minutes?
(a) 80 gallons
(b) 160 gallons
(c) 800 gallons
(d) 1650 gallons

93. Use the following data to answer this question: Ramu keeps track of the length of each fish that he catches. Below are the lengths in inches of the fish that he caught one day:
   12, 13, 8, 10, 8, 9, 17
What is the median fish length that Ramu caught that day?
(a) 8 inches
(b) 10 inches
(c) 11 inches
(d) 12 inches

94. If it takes two workers, working at the same speed, 2 hours 40 minutes to complete a particular task, about how long will it take one worker to complete the same task alone?
(a) 1 hour 20 minutes
(b) 4 hours 40 minutes
(c) 5 hours
(d) 5 hours 20 minutes

95. A snack machine accepts only quarter coins (25 paise coin). Candy bars cost 25 paise, packages of peanuts cost 75 paise, and cans of cola cost 50 paise. How many quarters are needed to buy two candy bars, one package of peanuts, and one can of cola?
(a) 8 quarters
(b) 7 quarters
(c) 6 quarters
(d) 5 quarters

96. A student starts for his school at 9.40 A.M. and comes back home at 4.15 P.M. Determine the time spent by him in the school if half an hour is spent in coming and going.
(a) 5 hours 25 minutes
(b) 6 hours 5 minutes
(c) 13 hours 55 minutes
(d) 4 hours 55 minutes

1000 MATH PROBLEMS >>> Miscellaneous Math

SET 7

97. An elevator sign reads "Maximum weight 300 kgs." Which of the following may ride the elevator?
    (a) three people: one weighing 98 kgs, one weighing 85 kgs, one weighing 100 kgs
    (b) one person weighing 82 kgs with a load weighing 250 kgs
    (c) one person weighing 65 kgs with a load weighing 243 kgs
    (d) three people: one weighing 110 kgs, one weighing 101 kgs, one weighing 98 kgs

98. Anju was hired to teach 3 identical math courses, which entailed being present in the classroom 48 hours altogether. At Rs. 35 per class hour, how much did Anju earn for teaching 1 course?
    (a) Rs. 105
    (b) Rs. 560
    (c) Rs. 840
    (d) Rs. 1,680

99. Anita and Manju got summer jobs at the Dairy and were supposed to work 15 hours per week each for 8 weeks. During that time, Manju was ill for one week and Anita took her shifts. How many hours did Anita work during the 8 weeks?
    (a) 120 hours
    (b) 135 hours
    (c) 150 hours
    (d) 185 hours

100. 11376 trees are planted in 48 gardens. Find the number of trees in each garden if all the gardens have the same number of trees.
     (a) 237 trees
     (b) 327 trees
     (c) 732 trees
     (d) 273 trees

101. When simplifying a word problem, what is the best way to write "Burt is 46 years old"?
     (a) Burt is 46
     (b) 46 = B
     (c) B = 46
     (d) B + 46

102. Dinesh went fishing six days in the month of June. He caught 11, 4, 0, 5, 4, and 6 fishes respectively. On the days that Dinesh fished, what was his average catch?
     (a) 4 fishes
     (b) 5 fishes
     (c) 6 fishes
     (d) 7 fishes

1000 MATH PROBLEMS >>> *Miscellaneous Math*

Answer question 103 on the basis of the following paragraph.

Basic cable television service, which includes 16 channels, cost Rs. 15 a month. The initial labor fee to install the service is Rs. 25. A Rs. 65 deposit is required but will be refunded within two years if the customer's bills are paid in full. Other cable services may be added to the basic service: the movie channel service is Rs. 9.40 a month; the news channels are Rs. 7.50 a month; the arts channels are Rs. 5.00 a month; the sports channels are Rs 4.80 a month.

103. A customer's first bill after having cable television installed totaled Rs 112.50. This customer chose basic cable and one additional cable service. Which additional service was chosen?
   (a) the news channels
   (b) the movie channels
   (c) the arts channels
   (d) the sports channels

104. An Army food supply truck can carry 3 tons. A breakfast ration weighs 200 gms, and the other two daily meals weigh 275 gms each. Assuming each soldier gets 3 meals per day, on a ten-day trip, how many soldiers can be supplied by one truck?
   (a) 100 soldiers
   (b) 150 soldiers
   (c) 200 soldiers
   (d) 320 soldiers

105. Manu arrived at work at 8:14 am. and Kiran arrived at 9:12 am. How long had Manu been at work when Kiran got there?
   (a) 1 hour 8 minutes
   (b) 1 hour 2 minutes
   (c) 58 minutes
   (d) 30 minutes

106. A clerk can process 26 forms per hour. If 5,600 forms must be processed in an 8-hour day, how many clerks must you hire for that day?
   (a) 24 clerks
   (b) 25 clerks
   (c) 26 clerks
   (d) 27 clerks

107. On the same latitude, Company E travels east at 35 miles per hour and Company F travels west at 15 miles per hour. If the two companies start out 2,100 miles apart, how long will it take them to meet?
   (a) 42 hours
   (b) 60 hours
   (c) 105 hours
   (d) 140 hours

108. What is the best way to simplify the following sentence to make it easier to work with? Ravi had three apples and ate one.
   (a) $R = 3 - 1$
   (b) $3 - 2 = R$
   (c) $R = 2$
   (d) $3R - 2$

109. A uniform requires 4 metres of cloth. To produce uniforms for 84,720 troops, how much cloth is required?

(a) 21,180 square metres
(b) 21,880 square metres
(c) 338,880 square metres
(d) 340,880 square metres

110. A law enforcement agency receives a report of a drunk driver on the roadway on August 3 at 10:42 p.m. and another similar report at 1:19 a.m. on August 4. How much time has elapsed between reports?

(a) 1 hour 37 minutes
(b) 2 hours 23 minutes
(c) 2 hours 37 minutes
(d) 3 hours 23 minutes

111. Studies have shown that automatic sprinkler systems save about Rs 5,700 in damages per fire in stores and offices. If a particular community has an average of 14 store and office fires every year, about how much money is saved each year if these buildings have sprinkler systems?

(a) Rs 28,500
(b) Rs 77,800
(c) Rs 79,800
(d) Rs 87,800

112. A certain commercial weight-loss clinic has devised a table of average desirable weights, based on heights. Using the information below, estimate the average desirable weight of a person who is 5 feet 5 inches tall.

Height	Weight
5'	80 kgs
6'	100 kgs

(a) 125 kgs
(b) 130 kgs
(c) 105 kgs
(d) 140 kgs

SET 8

113. Lata buys three puppies at the Friends Kennel for a total cost of Rs. 70. Two of the puppies are on sale for Rs. 15 a piece. How much does the third puppy cost?

(a) Rs. 55
(b) Rs. 40
(c) Rs. 30
(d) Rs. 25

Answer question 114 on the basis of the following table.

Monthly Taxes

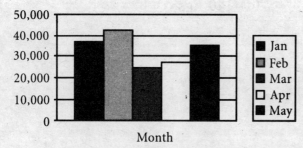

114. What were the total taxes collected for January, February, and April?

(a) Rs. 78,000
(b) Rs. 98,000
(c) Rs. 105,000
(d) Rs. 115,000

115. If Laxmi purchases an item that costs Rs 30 or less she will pay with cash.

If Laxmi purchases an item that costs between Rs 30 and Rs 70 she will pay with a cheque.

If Laxmi purchases an item that costs Rs 70 or greater, she will use a credit card.

If Laxmi recently paid for a certain item using a cheque, which of the following statements could be true?

(a) The item cost Rs 80
(b) If the item had cost Rs 20 more, she would have paid with cash
(c) The item cost at least Rs 70
(d) The item cost more than Rs 25

116. What is the next number in the series below?

3   16   6   12   12   8   —

(a) 4
(b) 15
(c) 20
(d) 24

117. Five people in Sonja's office are planning a party. Sonja will buy a loaf of bread (Rs. 3 a loaf) and a platter of cold cuts (Rs. 23). Babita will buy the soda (Rs. 1 per person) and two boxes of crackers (Rs. 2 per box). Mani and Ravi will split the cost of two packages of Cheese Doodles (Rs. 1 per package). Danny will supply a package of 5 paper plates (Rs. 4 per package). How much more will poor Sonja spend than the rest of the office put together?

(a) Rs. 14
(b) Rs. 13
(c) Rs. 12
(d) Rs. 11

1000 MATH PROBLEMS >>> Miscellaneous Math

118. Mr Ramakanth has inherited some musical instruments from his father. They are:
- 1 violin valued at   Rs. 3,500
- 2 violin bows, each valued at   Rs. 850
- 2 music stands, each valued at   Rs. 85
- 1 cello valued at   Rs. 2,300

In addition, Mr. Ramakanth's father has left him a watch valued at Rs. 250, and some old sheet music valued at Rs. 85 total. What is the value of Mr. Ramakanth's inheritance?

(a) Rs. 6,735
(b) Rs. 7,070
(c) Rs. 7,670
(d) Rs. 8,005

119. Six passengers Ashi, Rashmi, Ravi, Manish, Soni and Bharat, decide to go on a trip in a small plane that carries 6 people plus the pilot. At the last minute, Bharat and Soni decide to drive instead. Soni's friend, Rama decides she'd like to take Soni's place. This annoys Ashi who decides to stay at home. How many people, including the pilot, take the airoplane trip?

(a) 9 people
(b) 6 people
(c) 5 people
(d) 4 people

120. If a population of yeast cells grows from 10 to 320 in a period of 5 hours, what is the rate of growth?

(a) It doubles its numbers every hour.
(b) It triples its numbers every hour.
(c) It doubles its numbers every two hours.
(d) It triples its numbers every two hours.

121. The number of red blood corpuscles in one cubic millimetre is about 5,000,000 and the number of white blood corpuscles in one cubic millimetre is about 8,000. What, then, is the ratio of white blood corpuscles to red blood corpuscles?

(a) 1 : 625
(b) 1 : 40
(c) 4 : 10
(d) 5 : 1,250

122. Girish has made a vow to jog for an average of one hour daily five days a week. He cut his workout short on Wednesday by 40 minutes but was able to make up 20 minutes on Thursday and 13 minutes on Friday. How many minutes of jogging did Girish lose for the week?

(a) 20 minutes
(b) 13 minutes
(c) 7 minutes
(d) 3 minutes

123. Toni carries about 300 people in her cab each month. If she drives 15 days each month, how many passengers does she average per day in a month?

    (a) 15 passengers
    (b) 20 passengers
    (c) 30 passengers
    (d) 32 passengers

124. Property on the shore of Lake Ganga is selling for Rs. 250 a front foot. If Mary buys a lot 300 feet wide, how much will she have to pay for it?

    (a) Rs. 12,500
    (b) Rs. 15,000
    (c) Rs. 75,000
    (d) Rs. 1,25,000

125. Joni is 5 feet 11 inches tall and Moni is 6 feet 5 inches tall. How much taller is Moni than Joni?

    (a) 1 foot 7 inches
    (b) 1 foot
    (c) 7 inches
    (d) 6 inches

126. Manju is 10 years older than Hira, who is 16. How old is Manju.

    (a) 6 years old
    (b) 20 years old
    (c) 26 years old
    (d) 30 years old

127. Which of the following numbers can be divided evenly by 19?

    (a) 54
    (b) 63
    (c) 76
    (d) 82

128. Find the mode of the following series of numbers:

    4 6 8 8 8 9 9 12

    (a) 7
    (b) 8
    (c) 9
    (d) 12

1000 MATH PROBLEMS >>> Miscellaneous Math

SET 9

129. Minu is 1 year 7 months old and Beena is 2 years 8 months old. How much older is Beena than Minu.

    (a) 1 year 1 month
    (b) 2 years
    (c) 1 month
    (d) 1 year 2 months

130. 5th graders Kara and Rani both have lemonade stands. Kara sells her lemonade at 50 paise glass and Rani sells her lemonade at 7 paise glass. Kara sold 17 glasses of lemonade today and Rani sold 14 glasses. Who made the most money and by what amount?

    (a) Kara, 200 paise
    (b) Rani, 200 paise
    (c) Kara, 85 paise
    (d) Rani, 1050 paise

131. A man earns Rs. 51,858.00 in a year. If he earns the same amount every month find his monthly Income

    (a) Rs. 4321.50
    (b) Rs. 3421.50
    (c) Rs. 2314.00
    (d) Rs. 1728.60

132. If Rita can run around the garden 5 times in 20 minutes, how many times can she run around the garden in one hour?

    (a) 10
    (b) 15
    (c) 50
    (d) 100

133. Which of the following number sentences is true?

    (a) 4 feet > 3 feet
    (b) 7 feet < 6 feet
    (c) 5 feet > 6 feet
    (d) 3 feet < 2 feet

134. Which of the following is a prime (not a composite) number?

    (a) 4
    (b) 8
    (c) 11
    (d) 27

135. Which of the following is the best simplification of the following sentence? Soni is ten years older than Rama.

    (a) 10 + S = R
    (b) S + R = 10
    (c) R − 10 = S
    (d) S = R + 10

136. Which of the following is a synonym for "average"?

    (a) mode
    (b) quotient
    (c) mean
    (d) median

137. At the movies, Lalith bought food for herself and her friend Ram, including : 1 box of popcorn to share at Rs. 5 a box, 1 box of Junior Mints for each of them at Rs. 2 a box, and 1 soft drink for each at Rs. 3 for each. Ram bought a ticket at Rs. 7 for each. Who spent the most money and by how much?

    (a) Ram by Rs. 3
    (b) Ram by Rs. 7
    (c) Lalith by Rs. 1
    (d) Lalith by Rs. 2

1000 MATH PROBLEMS >>> *Miscellaneous Math*

138. One colony of bats consumes 36 tons of mosquitoes per year. At that rate, how many kgs of mosquitoes does the same colony consume in a month?

(a) 36,000 kgs
(b) 12,000 kgs
(c) 6,000 kgs
(d) 3,000 kgs

139. In 1995, the number of insects on earth was estimated at 10^{18}. How many insects were there?

(a) 10 × 10 eighteen times
(b) 10 + 18 ten times
(c) 10 million × 18 million
(d) 18 × 18 ten times

140. How many acres are contained in a parcel 121 feet wide and 240 yards deep? (1 acre = 43,560 square feet).

(a) 1 acre
(b) $1\frac{1}{2}$ acres
(c) 2 acres
(d) $2\frac{1}{2}$ acres

141. Land consisting of a quarter section is sold for Rs. 1,850 per acre (1 quarter section = 160 acres). The total sale price is

(a) Rs. 296,000
(b) Rs. 592,000
(c) Rs. 1,184,000
(d) Rs. 1,850,000

142. The planet Pluto is estimated at a mean distance of 3,666 million miles from the sun. The planet Mars is estimated at a mean distance of 36 million miles from the sun. How much closer to the sun is Mars than Pluto?

(a) 36,300,000 million miles
(b) 36,300 million miles
(c) 3,630 million miles
(d) 363 million miles

143. A rectangular tract of land measures 860 feet by 560 feet. Approximately how many acres is this? (1 acre = 43,560 square feet.)

(a) 12.8 acres
(b) 11.05 acres
(c) 10.5 acres
(d) 8.06 acres

144. A dormitory now houses 30 men and allows 42 square feet of space per man. If five more men are put into this dormitory, how much less space will each man have?

(a) 5 square feet
(b) 6 square feet
(c) 7 square feet
(d) 8 square feet

1000 MATH PROBLEMS >>> Miscellaneous Math

SET 10

145. To lower a fever of 105 degrees, ice packs are applied for 1 minute and then removed for 5 minutes before being applied again. Each application lowers the fever by half a degree. How long will it take to lower the fever to 99 degrees?
- (a) 1 hour
- (b) 1 hour and 12 minutes
- (c) 1 hour and 15 minutes
- (d) 1 hour and 30 minutes

146. What is the median of the following group of numbers?
6 8 10 12 14 16 18
- (a) 11
- (b) 12
- (c) 13
- (d) 14

147. What is the median of the following group of numbers?
10 12 14 16 18 20
- (a) 14
- (b) 15
- (c) 16
- (d) 20

148. $4\sqrt{3} - 2\sqrt{3} =$
- (a) $2\sqrt{3}$
- (b) $2-\sqrt{3}$
- (c) 2
- (d) 6

149. What is another way to write $3\sqrt{12}$?
- (a) $12\sqrt{3}$
- (b) $6\sqrt{3}$
- (c) $2\sqrt{10}$
- (d) 18

150. Which of the following numbers is evenly divisible by 3?
- (a) 235
- (b) 236
- (c) 237
- (d) 238

151. Simplify the following: Rinku has three times the number of tennis trophies Avinash has.
- (a) R = A + 3
- (b) A = R - 3
- (c) R = 3A
- (d) A = 3R

152. Write ten thousand four hundred forty-seven in numerals.
- (a) 10,499,047
- (b) 104,447
- (c) 10,447
- (d) 1,047

153. Write ten million forty-three thousand seven hundred and three in numerals.
- (a) 143,703
- (b) 1,043,703
- (c) 10,043,703
- (d) 10,430,703

154. John has started an egg farm. His free-range chickens produce 480 eggs per day, and his eggs sell for Rs. 2.00 a dozen. How much does John make on eggs per week? (1 week = 7 days).
- (a) Rs. 480
- (b) Rs. 500
- (c) Rs. 560
- (d) Rs. 600

1000 MATH PROBLEMS >>> Miscellaneous Math

155. When he got out of college, Aman started gaining weight at a steady rate. In ten years he gained 60 kgs. How much weight did Aman gain per year?
    (a) 30 kgs
    (b) 15 kgs
    (c) 6 kgs
    (d) 10 kgs

156. Punit could not get serious about working and changed jobs 27 times between 1991 and 1993. On average, how many jobs did Punit hold in each of those three years?
    (a) 12 jobs
    (b) 9 jobs
    (c) 6 jobs
    (d) 4 jobs

Answer question 157 on the basis of the following table.

Production of Farm-it Tractors for the month of April

Factory	April Output
Doves	450
Hindustan	425
Lunar	
Ambani	345
TOTAL	1780

157. What was Lunar's production in the month of April?
    (a) 345
    (b) 415
    (c) 540
    (d) 560

158. Fifty-four students are to be separated into six groups of equal size. How many students are in each group?
    (a) 8 students
    (b) 9 students
    (c) 10 students
    (d) 12 students

Answer question 159 on the basis of the table below

Suraj's Bird-watching Project

Day	Number of Raptors Seen
Monday	
Tuesday	7
Wednesday	12
Thursday	11
Friday	4
MEAN	8

159. The table above shows the data Suraj collected while watching birds for one week. How many raptors did Suraj see on Monday?
    (a) 6 raptors
    (b) 7 raptors
    (c) 8 raptors
    (d) 10 raptors

Answer question 160 on the basis of the table below.

Distance travelled from Delhi with Respect to Time

Time (hours)	Distance from Delhi (miles)
1	60
2	120
3	180
4	240

160. A train moving at a constant speed leaves Delhi for Patna at time $t = 0$. If Patna is 2,000 miles from Delhi, which of the following equations describes the distance from Patna at any time t?
    (a) $D(t) = 60t - 2000$
    (b) $D(t) = 60t$
    (c) $D(t) = 2000 - 60t$
    (d) $D(t) = 2000 + 60t$

SECTION

2

FRACTIONS

The following 10 sets of fraction problems will provide you with exercises in how to convert fractions and how to do arithmetic problems that involve fractions. (Sections 3 and 4 deal with decimals and percentages, which are also actually fractions, but are called something else for clarity.) In order to understand arithmetic in general, it is important to practice and become comfortable with fractions and how they work.

You'll start off with questions that just deal with numbers. After you've had a chance to practice your basic fraction skills, you can move on to some word problems involving fractions.

1000 MATH PROBLEMS >>> Fractions

SET 11

161. Name the fraction that indicates the shaded part of the figure below.

(a) $\frac{1}{4}$

(b) $\frac{1}{2}$

(c) $\frac{2}{3}$

(d) $\frac{3}{4}$

162. Name the fraction that indicates the shaded part of the figure below.

(a) $\frac{2}{5}$

(b) $\frac{3}{5}$

(c) $\frac{5}{3}$

(d) $\frac{5}{2}$

163. $\frac{7}{8} - \frac{3}{5} =$

(a) $\frac{11}{40}$

(b) $1\frac{1}{3}$

(c) $\frac{1}{10}$

(d) $1\frac{19}{40}$

164. $2\frac{1}{4} + 4\frac{5}{8} + \frac{1}{2} =$

(a) $6\frac{7}{8}$

(b) $7\frac{1}{4}$

(c) $7\frac{3}{8}$

(d) $7\frac{3}{4}$

165. $3\frac{7}{10} - 2\frac{3}{8} =$

(a) $1\frac{13}{40}$

(b) $1\frac{7}{20}$

(c) $1\frac{11}{18}$

(d) $2\frac{1}{80}$

166. $\frac{1}{6} + \frac{7}{12} + \frac{2}{3} =$

(a) $\frac{10}{24}$

(b) $2\frac{1}{6}$

(c) $1\frac{5}{6}$

(d) $1\frac{5}{12}$

167. $4\frac{1}{3} + 3\frac{3}{10} =$

(a) $7\frac{2}{15}$

(b) $7\frac{4}{13}$

(c) $7\frac{2}{3}$

(d) $7\frac{19}{30}$

168. $3\frac{9}{16} - 1\frac{7}{8} =$

(a) $1\frac{11}{16}$

(b) $2\frac{1}{8}$

(c) $2\frac{1}{4}$

(d) $2\frac{5}{16}$

1000 MATH PROBLEMS >>> Fractions

169. $4\frac{2}{5} + 3\frac{1}{2} + \frac{3}{8} =$

   (a) $7\frac{3}{20}$

   (b) $7\frac{2}{5}$

   (c) $8\frac{11}{40}$

   (d) $8\frac{7}{8}$

170. $\frac{5}{12} - \frac{3}{8} =$

   (a) $\frac{1}{10}$

   (b) $\frac{1}{24}$

   (c) $\frac{5}{48}$

   (d) $\frac{19}{24}$

171. $56\frac{3}{8} - 10\frac{5}{6} =$

   (a) $46\frac{1}{7}$

   (b) $46\frac{13}{14}$

   (c) $45\frac{1}{3}$

   (d) $45\frac{13}{24}$

172. $-\frac{3}{10} \div -\frac{1}{5} =$

   (a) $1\frac{1}{2}$

   (b) $\frac{2}{3}$

   (c) $-\frac{3}{50}$

   (d) $\frac{3}{50}$

173. $30 \div 2\frac{1}{2} =$

   (a) $\frac{1}{15}$

   (b) 15

   (c) 12

   (d) 75

174. $\frac{1}{3} \div \frac{2}{7} =$

   (a) $2\frac{4}{5}$

   (b) $1\frac{1}{6}$

   (c) $2\frac{1}{7}$

   (d) $1\frac{1}{5}$

175. $\frac{7}{8} - \frac{3}{5} =$

   (a) $\frac{11}{40}$

   (b) $1\frac{1}{3}$

   (c) $\frac{1}{10}$

   (d) $1\frac{19}{40}$

176. $4\frac{1}{5} + 1\frac{2}{5} + 3\frac{3}{10} =$

   (a) $9\frac{1}{10}$

   (b) $8\frac{9}{10}$

   (c) $8\frac{4}{5}$

   (d) $8\frac{6}{15}$

1000 MATH PROBLEMS >>> Fractions

SET 12

177. $76\frac{1}{2} + 11\frac{5}{6} =$

(a) $87\frac{1}{2}$

(b) $88\frac{1}{3}$

(c) $88\frac{5}{6}$

(d) $89\frac{1}{6}$

178. $20\frac{5}{7} - 15\frac{1}{7} =$

(a) $5\frac{4}{14}$

(b) $5\frac{3}{7}$

(c) $5\frac{4}{7}$

(d) $5\frac{6}{7}$

179. $43\frac{2}{3} + 36\frac{3}{9} =$

(a) 100

(b) 90

(c) 80

(d) 70

180. $\frac{5}{8} - \frac{1}{3} =$

(a) $3\frac{1}{4}$

(b) $\frac{4}{24}$

(c) $3\frac{2}{5}$

(d) $\frac{7}{24}$

181. $\frac{6}{7} + \frac{4}{5} =$

(a) $\frac{25}{44}$

(b) $\frac{35}{49}$

(c) $\frac{14}{33}$

(d) $\frac{58}{35}$

182. $\frac{3}{7} \times \frac{7}{3} =$

(a) $1\frac{1}{3}$

(b) 1

(c) $\frac{3}{21}$

(d) $\frac{1}{21}$

183. $-\frac{1}{4} \div -\frac{1}{8} =$

(a) $-\frac{1}{2}$

(b) $\frac{1}{2}$

(c) 2

(d) -2

184. $3\frac{3}{4} \times \frac{4}{5} =$

(a) $\frac{3}{4}$

(b) 3

(c) $3\frac{3}{4}$

(d) $4\frac{11}{16}$

185. $\frac{5}{8} \div 3 =$

(a) $\frac{5}{24}$

(b) $\frac{3}{8}$

(c) $1\frac{7}{8}$

(d) $\frac{3}{24}$

1000 MATH PROBLEMS >>> *Fractions*

186. $2\frac{1}{4} \div 2\frac{4}{7} =$
    (a) $\frac{9}{14}$
    (b) $\frac{7}{8}$
    (c) $1\frac{2}{7}$
    (d) $5\frac{11}{14}$

187. $\frac{3}{10} \times \frac{4}{5} \times \frac{1}{2} =$
    (a) $\frac{3}{4}$
    (b) $\frac{12}{25}$
    (c) $\frac{13}{27}$
    (d) $\frac{3}{25}$

188. $1\frac{1}{2} \div 1\frac{5}{13} =$
    (a) $1\frac{3}{10}$
    (b) $1\frac{1}{12}$
    (c) $2\frac{1}{13}$
    (d) $3\frac{9}{10}$

189. $2\frac{1}{3} \times 1\frac{1}{14} \times 1\frac{4}{5} =$
    (a) $1\frac{7}{18}$
    (b) $2\frac{1}{2}$
    (c) $3\frac{6}{7}$
    (d) $4\frac{1}{2}$

190. $\frac{2}{5} \times \frac{3}{7} =$
    (a) $\frac{6}{35}$
    (b) $\frac{14}{15}$
    (c) $\frac{5}{12}$
    (d) $\frac{29}{35}$

191. $2\frac{1}{4} \div \frac{2}{3} =$
    (a) $\frac{8}{27}$
    (b) $1\frac{1}{2}$
    (c) $3\frac{3}{8}$
    (d) $3\frac{1}{2}$

192. $\frac{2}{3} \div \frac{5}{12} =$
    (a) $1\frac{3}{5}$
    (b) $1\frac{5}{18}$
    (c) $1\frac{7}{36}$
    (d) $1\frac{5}{6}$

SET 13

193. $4 \times \frac{1}{3} =$

(a) $1\frac{1}{3}$

(b) $1\frac{1}{2}$

(c) $2\frac{1}{4}$

(d) $2\frac{3}{4}$

194. $6 \times \frac{2}{3} =$

(a) $\frac{1}{9}$

(b) $\frac{2}{3}$

(c) $2\frac{2}{3}$

(d) 4

195. $\frac{1}{4} \div 2\frac{4}{7}$

(a) $\frac{9}{14}$

(b) $\frac{7}{72}$

(c) $1\frac{2}{7}$

(d) $\frac{9}{14}$

196. $2\frac{5}{8} \div \frac{1}{3} =$

(a) $8\frac{1}{3}$

(b) $7\frac{7}{8}$

(c) $5\frac{11}{24}$

(d) $\frac{7}{8}$

197. $1\frac{1}{2} \div 1\frac{5}{13} =$

(a) $1\frac{3}{10}$

(b) $1\frac{1}{12}$

(c) $2\frac{1}{13}$

(d) $3\frac{9}{10}$

198. $2\frac{1}{3} \times 1\frac{1}{14} \times 1\frac{4}{5} =$

(a) $1\frac{7}{18}$

(b) $2\frac{1}{2}$

(c) $3\frac{6}{7}$

(d) $4\frac{1}{2}$

199. $7\frac{3}{5} \div \frac{1}{4} =$

(a) $24\frac{2}{7}$

(b) $27\frac{3}{8}$

(c) $30\frac{2}{5}$

(d) $33\frac{1}{2}$

200. $\frac{3}{5} \times 1\frac{1}{3} =$

(a) $\frac{9}{14}$

(b) $\frac{12}{15}$

(c) $\frac{14}{22}$

(d) $\frac{11}{9}$

1000 MATH PROBLEMS >>> Fractions

201. $-\frac{5}{3} - \frac{1}{3} =$
   (a) $\frac{4}{3}$
   (b) $-\frac{4}{3}$
   (c) 2
   (d) -2

202. $\frac{7}{8} \times \frac{1}{4} =$
   (a) $4\frac{1}{2}$
   (b) $\frac{7}{32}$
   (c) $3\frac{1}{8}$
   (d) $\frac{2}{7}$

203. $7 \div \frac{3}{8} =$
   (a) $18\frac{2}{3}$
   (b) $12\frac{3}{8}$
   (c) $14\frac{5}{6}$
   (d) $10\frac{4}{5}$

204. $6\frac{2}{9} - \frac{1}{6} =$
   (a) $6\frac{1}{18}$
   (b) $6\frac{5}{27}$
   (c) $6\frac{1}{3}$
   (d) $5\frac{8}{9}$

205. $9\frac{3}{7} + 4\frac{2}{5} =$
   (a) $13\frac{5}{12}$
   (b) $13\frac{6}{8}$
   (c) $13\frac{29}{35}$
   (d) $13\frac{37}{52}$

206. $3\frac{5}{6} \times 4\frac{2}{3} =$
   (a) $17\frac{8}{9}$
   (b) $12\frac{7}{18}$
   (c) $16\frac{2}{3}$
   (d) $13\frac{3}{5}$

207. Which of the following is between $\frac{1}{3}$ and $\frac{1}{4}$?
   (a) $\frac{1}{5}$
   (b) $\frac{2}{3}$
   (c) $\frac{2}{5}$
   (d) $\frac{2}{7}$

208. Change $\frac{17}{3}$ to a mixed number.
   (a) $6\frac{1}{3}$
   (b) $5\frac{2}{3}$
   (c) $5\frac{1}{3}$
   (d) $4\frac{3}{4}$

SET 14

209. Which of the following is the equivalent of $\frac{13}{25}$?

(a) 0.38
(b) 0.4
(c) 0.48
(d) 0.52

210. Which of the following has the greatest value?

(a) $\frac{7}{8}$
(b) $\frac{3}{4}$
(c) $\frac{2}{3}$
(d) $\frac{5}{6}$

211. Which of the following numbers is the smallest?

(a) $\frac{6}{10}$
(b) $\frac{8}{15}$
(c) $\frac{33}{60}$
(d) $\frac{11}{20}$

212. Which of the following diameters is the smallest?

(a) $\frac{17}{20}$ inches
(b) $\frac{3}{4}$ inches
(c) $\frac{5}{6}$ inches
(d) $\frac{7}{10}$ inches

213. What is the reciprocal of $3\frac{7}{8}$?

(a) $\frac{31}{8}$
(b) $\frac{8}{31}$
(c) $\frac{8}{21}$
(d) $-\frac{31}{8}$

214. What is the reciprocal of $3\frac{3}{4}$?

(a) $\frac{4}{15}$
(b) $\frac{15}{4}$
(c) $\frac{14}{5}$
(d) $\frac{5}{14}$

215. Change $\frac{160}{40}$ to a whole number.

(a) 16
(b) 10
(c) 8
(d) 4

216. Change this improper fraction to a mixed number: $\frac{15}{2}$.

(a) 8
(b) $7\frac{1}{2}$
(c) 7
(d) $6\frac{1}{2}$

217. Change this mixed number to an improper fraction: $5\frac{1}{2}$

(a) $\frac{11}{2}$
(b) $\frac{10}{2}$
(c) $\frac{7}{2}$
(d) $\frac{5}{2}$

1000 MATH PROBLEMS >>> Fractions

218. $\frac{2}{3} - \frac{1}{5} =$

   (a) $\frac{7}{15}$

   (b) $\frac{2}{5}$

   (c) $\frac{5}{12}$

   (d) $\frac{3}{8}$

219. Which of the following is an improper fraction?

   (a) $\frac{22}{60}$

   (b) $\frac{66}{22}$

   (c) $\frac{90}{100}$

   (d) $\frac{1000}{2600}$

220. Manish has finished 35 out of 45 of his test questions. Which of the following fractions of the test does he has left?

   (a) $\frac{2}{9}$

   (b) $\frac{7}{9}$

   (c) $\frac{4}{5}$

   (d) $\frac{3}{5}$

221. Jona gave $\frac{1}{2}$ of her sandwich to Mona at lunch-time, and ate $\frac{1}{3}$ of it herself. How much of the sandwich did she has left?

   (a) $\frac{1}{6}$

   (b) $\frac{3}{5}$

   (c) $\frac{4}{5}$

   (d) $\frac{5}{6}$

222. Kiran is buying fabric for new curtains. There are three windows, each 35 inches wide. Kiran needs to buy fabric equal to $2\frac{1}{2}$ times the total width of the windows. How much fabric should he buy?

   (a) $262\frac{1}{2}$ inches

   (b) $175\frac{1}{3}$ inches

   (c) $210\frac{3}{4}$ inches

   (d) $326\frac{1}{4}$ inches

223. Raman made sweet for dinner last night. He and his family ate $\frac{2}{3}$ of it and saved the rest. The next day, Raman ate $\frac{1}{2}$ of the remainder for lunch. What fraction of the original sweet is left?

   (a) $\frac{1}{5}$

   (b) $\frac{1}{6}$

   (c) $\frac{1}{7}$

   (d) $\frac{1}{8}$

224. Monica needs $\frac{5}{8}$ of a cup of diced onion for a recipe. After chopping all the onions she has, $\frac{3}{5}$ of a cup of chopped onions. How much more chopped onions does she need?

   (a) $\frac{1}{8}$ of a cup

   (b) $\frac{1}{5}$ of a cup

   (c) $\frac{1}{40}$ of a cup

   (d) $\frac{1}{60}$ of a cup

1000 MATH PROBLEMS >>> Fractions

SET 15

225. Ram has $5\frac{1}{2}$ kgs of sugar. He wants to make cookies for his son's kindergarden class. The cookie recipe calls for $\frac{2}{3}$ kgs of sugar per dozen cookies. How many dozen cookies can he make?

(a) $6\frac{1}{3}$ dozen cookies

(b) $7\frac{1}{5}$ dozen cookies

(c) $8\frac{1}{4}$ dozen cookies

(d) $9\frac{1}{2}$ dozen cookies

226. For the company's third anniversary, the caterer provided 3 one-kgs of cheese. At the end of the party, there were $\frac{3}{5}$ kgs of swiss, $\frac{4}{7}$ kgs of Vermont cheddar, and $\frac{5}{8}$ kgs of feta cheese left. What fraction of the original three kgs was left after the party?

(a) $1\frac{123}{280}$ kgs of cheese

(b) $1\frac{223}{280}$ kgs of cheese

(c) $1\frac{283}{270}$ kgs of cheese

(d) $1\frac{393}{290}$ kgs of cheese

227. Vanita is making a mosaic. Each tiny piece of glass in the artwork is $1\frac{1}{4}$ inch by $1\frac{3}{8}$ inch. What is the area of each piece?

(a) $1\frac{23}{32}$ square inches

(b) $1\frac{21}{22}$ square inches

(c) $1\frac{23}{25}$ square inches

(d) $1\frac{29}{31}$ square inches

228. Dave has $4\frac{1}{2}$ kgs of taco chips for tonight's poker game. There are seven players (including Dave). If they all have equal portions of the chips, how many kgs of chips does each player get?

(a) 2 kgs

(b) $\frac{1}{2}$ of a kg

(c) $\frac{3}{4}$ of a kg

(d) $\frac{9}{14}$ of a kg

229. Maya has $17\frac{3}{4}$ feet of wallpaper border. Each wall of her bathroom is nine feet long. How much more wallpaper border does Maya need?

(a) $17\frac{3}{4}$ feet

(b) $16\frac{1}{2}$ feet

(c) $18\frac{1}{4}$ feet

(d) $19\frac{1}{2}$ feet

1000 MATH PROBLEMS >>> Fractions

230. A recipe calls for all the liquid ingredients to be mixed together: $2\frac{1}{4}$ cups of water, $4\frac{5}{8}$ cups of chicken stock, and $\frac{1}{2}$ cup of honey. How many cups of liquid are in the recipe?

(a) $6\frac{7}{8}$ cups

(b) $7\frac{1}{4}$ cups

(c) $7\frac{3}{8}$ cups

(d) $7\frac{3}{4}$ cups

231. A loaf of bread has 35 slices. Anju eats 8 slices, Beena eats 6 slices, Sanju eats 5, and Dinesh eats 9 slices. What fraction of the loaf is left?

(a) $\frac{2}{11}$

(b) $\frac{1}{9}$

(c) $\frac{2}{7}$

(d) $\frac{1}{5}$

232. Manisha wants to run $2\frac{1}{3}$ miles every day. Today she has gone $\frac{7}{8}$ mile. How much farther does she has to go?

(a) $1\frac{11}{24}$ miles

(b) $1\frac{1}{3}$ miles

(c) $1\frac{41}{50}$ miles

(d) $1\frac{307}{308}$ miles

233. Ribbon in a craft store costs Rs. 0.75 per yard. Vernon needs to buy $7\frac{1}{3}$ yards. How much will it cost?

(a) Rs. 7.33

(b) Rs. 6.95

(c) Rs. 5.50

(d) Rs. 4.25

234. Leena needs to read 14 pages for her History class, 26 pages for English, 12 pages for Civics, and 28 pages for Biology. She has read $\frac{1}{6}$ of the entire number of pages. How many pages has she read?

(a) 80 pages

(b) $13\frac{1}{3}$ pages

(c) $48\frac{1}{2}$ pages

(d) 17 pages

235. Toni has to write a $5\frac{1}{2}$ page paper. He's finished $3\frac{1}{3}$ pages. How many pages are he has left to write?

(a) $1\frac{3}{5}$ pages

(b) $1\frac{7}{8}$ pages

(c) $2\frac{2}{3}$ pages

(d) $2\frac{1}{6}$ pages

236. Mona made Rs. 331.01 last week. She worked $39\frac{1}{2}$ hours. What is her hourly wage?

(a) Rs. 8.28

(b) Rs. 8.33

(c) Rs. 8.38

(d) Rs. 8.43

1000 MATH PROBLEMS >>> Fractions

237. Veena ate $\frac{3}{7}$ of a chocolate chip cookie; Aruna ate $\frac{1}{3}$ of the same cookie. How much of the cookie is left?

(a) $\frac{1}{3}$ cookie

(b) $\frac{3}{7}$ cookie

(c) $\frac{7}{10}$ cookie

(d) $\frac{5}{21}$ cookie

238. Manju has worked $6\frac{5}{8}$ hours of her regular 8-hour day. How many more hours must she work?

(a) $1\frac{1}{2}$ hours

(b) $1\frac{3}{8}$ hours

(c) $2\frac{1}{4}$ hours

(d) $1\frac{1}{4}$ hours

239. Isha has read $\frac{3}{5}$ of the novel assigned for her English class. The novel is 360 pages long. How many pages has she read?

(a) 216 pages

(b) 72 pages

(c) 300 pages

(d) 98 pages

240. Ganesh ran $7\frac{1}{8}$ miles on Monday, $5\frac{2}{3}$ miles on Tuesday, $6\frac{2}{7}$ miles on Wednesday, $7\frac{1}{2}$ miles on Thursday, $5\frac{1}{4}$ miles on Friday and $6\frac{3}{5}$ miles on Saturday. On Sunday, he went to swimming. How many miles did he run this week?

(a) $29\frac{27}{100}$ miles

(b) $28\frac{319}{1000}$ miles

(c) $39\frac{33}{100}$ miles

(d) $38\frac{427}{1000}$ miles

SET 16

241. Mona ordered a claw hammer, four drill bits, a work light, a large clamp, two screwdrivers, seven toggle bolts, 16 two-penny nails, three paintbrushes, and a 48-inch level from a mail order house. So far, she has received the hammer, three drill bits, the level, one screwdriver, the clamp, and all the two-penny nails. What fraction of her order has she received?

(a) $\frac{1}{32}$

(b) $\frac{16}{23}$

(c) $\frac{23}{36}$

(d) $\frac{36}{23}$

242. Sanju makes Rs. 7.75 an hour. He worked $38\frac{1}{5}$ hours last week. How much money did he earn?

(a) Rs. 592.10
(b) Rs. 296.05
(c) Rs. 775.00
(d) Rs. 380.25

243. A lasagna recipe calls for $3\frac{1}{2}$ kgs of noodles. How many kgs of noodles are needed to make $\frac{1}{3}$ of a recipe?

(a) 1 kg
(b) $1\frac{1}{2}$ kgs
(c) $\frac{5}{6}$ kgs
(d) $1\frac{1}{6}$ kgs

244. A lasagna recipe requires $1\frac{1}{2}$ kgs of cheese. Approximately how many lasagnas can be made from a $20\frac{1}{3}$ kgs block of cheese?

(a) $13\frac{1}{2}$ recipes
(b) $20\frac{1}{3}$ recipes
(c) $10\frac{1}{5}$ recipes
(d) $25\frac{1}{4}$ recipes

245. For health reasons, Amir wants to drink eight glasses of water a day. He's already had six glasses. What fraction of eight glasses does Amir has left to drink?

(a) $\frac{1}{8}$
(b) $\frac{1}{6}$
(c) $\frac{1}{4}$
(d) $\frac{2}{6}$

246. Veena is writing a test to give to her History class. She wants the test to include 40 multiple-choice questions and 60 short answer questions. She has written 25 of the multiple-choice questions. What fraction of the total test has she written?

(a) $\frac{1}{4}$
(b) $\frac{5}{8}$
(c) $\frac{2}{3}$
(d) $\frac{5}{12}$

1000 MATH PROBLEMS >>> Fractions

247. Mahesh's car gets $14\frac{1}{3}$ miles per gallon. It's $58\frac{1}{2}$ miles from his home to work. How many gallons does Mahesh's car use on the way to work?
(a) $2\frac{9}{10}$ gallons
(b) $3\frac{1}{16}$ gallons
(c) $4\frac{7}{86}$ gallons
(d) $5\frac{3}{8}$ gallons

248. Uma needs 168 six-inch fabric squares to make a quilt top. She has 150 squares. What fraction of the total does she still need?
(a) $\frac{150}{168}$
(b) $\frac{9}{84}$
(c) $\frac{25}{28}$
(d) $\frac{3}{28}$

249. Raja wants to paint his living room ceiling red. His ceiling is $14\frac{1}{2}$ feet by $12\frac{1}{3}$ feet. One gallon of paint will cover 90 square feet. How many gallons of paint will he need?
(a) 1 gallon
(b) 2 gallons
(c) 3 gallons
(d) 4 gallons

250. The Garcias had $\frac{2}{5}$ of last night's meat loaf left over after dinner. Today, Uncle Khanna ate $\frac{1}{4}$ of these leftovers. How much of the original meat loaf is left?
(a) $\frac{3}{4}$
(b) $\frac{3}{10}$
(c) $\frac{3}{20}$
(d) $\frac{3}{5}$

251. Meena is a night security guard at the Art Museum. Each night, she is required to walk through each gallery once. The Museum contains 52 galleries. This night, Meena has walked through 16 galleries. What fraction of the total galleries has she already visited?
(a) $\frac{4}{13}$
(b) $\frac{1}{16}$
(c) $\frac{5}{11}$
(d) $\frac{3}{14}$

252. Arun has been ill and worked only $\frac{3}{4}$ of his usual 40 hour week. He makes Rs. 12.35 an hour. How much has he earned this week?
(a) Rs. 247.00
(b) Rs. 308.75
(c) Rs. 370.50
(d) Rs. 432.25

1000 MATH PROBLEMS >>> Fractions

253. A recipe calls for $\frac{1}{4}$ teaspoon of red pepper. How much red pepper would you need for half a recipe?
(a) $\frac{1}{10}$ teaspoon
(b) $\frac{1}{8}$ teaspoon
(c) $\frac{1}{6}$ teaspoon
(d) $\frac{1}{2}$ teaspoon

254. A recipe calls for $\frac{1}{4}$ teaspoon of red pepper. How much red pepper would you need for a double recipe?
(a) $\frac{1}{10}$ teaspoon
(b) $\frac{1}{8}$ teaspoon
(c) $\frac{1}{6}$ teaspoon
(d) $\frac{1}{2}$ teaspoon

255. Jona's lawn is 30 yards by 27 yards. Yesterday, Jona mowed $\frac{2}{3}$ of the lawn. How many square yards are left to be mowed today?
(a) 270 square yards
(b) 540 square yards
(c) 810 square yards
(d) 1080 square yards

256. A thirty-minute time slot on a television network contains 24 minutes of comedy and 6 minutes of commercials. What fraction of the program time is devoted to commercials?
(a) $\frac{1}{6}$ of the time
(b) $\frac{1}{4}$ of the time
(c) $\frac{1}{3}$ of the time
(d) $\frac{1}{5}$ of the time

SET 17

257. Manu has been on a diet and has lost ten kgs, or $\frac{1}{12}$ of his original weight. What was his original weight?
 (a) 120 kgs
 (b) 140 kgs
 (c) 160 kgs
 (d) 180 kgs

258. In a cashier contest, Ona packed $15\frac{1}{2}$ bags of groceries in 3 hours. How many bags did she average each hour?
 (a) $4\frac{1}{2}$ bags per hour
 (b) 5 bags per hour
 (c) $5\frac{1}{4}$ bags per hour
 (d) $5\frac{1}{6}$ bags per hour

259. It's $9\frac{3}{4}$ miles from Arun's house to his office. On Monday morning, he made it $\frac{1}{3}$ of the way before he ran out of gas. How far did he get?
 (a) $4\frac{1}{3}$ miles
 (b) $4\frac{1}{5}$ miles
 (c) $3\frac{1}{4}$ miles
 (d) $3\frac{1}{2}$ miles

260. At birth, Pinto weighed $6\frac{1}{2}$ kgs. At one year of age, he weighed $23\frac{1}{8}$ kgs. How much weight, in kgs, did he gain?
 (a) $16\frac{5}{8}$ kgs
 (b) $16\frac{7}{8}$ kgs
 (c) $17\frac{1}{6}$ kgs
 (d) $17\frac{3}{4}$ kgs

261. Monica wants to make muffins and needs $\frac{3}{4}$ cup of sugar. She discovers, however, that she has only $\frac{2}{3}$ cup of sugar. How much more sugar does she need?
 (a) $\frac{1}{12}$ cup
 (b) $\frac{1}{8}$ cup
 (c) $\frac{1}{6}$ cup
 (d) $\frac{1}{4}$ cup

262. How many inches are there in $3\frac{1}{3}$ yards?
 (a) 126 inches
 (b) 120 inches
 (c) 160 inches
 (d) 168 inches

1000 MATH PROBLEMS >>> Fractions

263. Anand's Candy Shop opened for business on Saturday with $22\frac{1}{4}$ kgs of fudge. During the day, they sold $17\frac{5}{8}$ kgs of fudge. How many kgs were left?
(a) $4\frac{1}{2}$ kgs
(b) $4\frac{5}{8}$ kgs
(c) $4\frac{7}{8}$ kgs
(d) $5\frac{3}{8}$ kgs

264. A child's swimming pool contains $20\frac{4}{5}$ gallons of water. If $3\frac{1}{3}$ gallons of water is splashed out of the pool while the children are playing, how many gallons of water is left?
(a) $16\frac{1}{15}$ gallons
(b) $16\frac{3}{5}$ gallons
(c) $17\frac{7}{15}$ gallons
(d) $17\frac{2}{3}$ gallons

265. During the month of May, $\frac{1}{6}$ of the buses in District A were in the garage for routine maintenance. In addition, $\frac{1}{8}$ of the buses were in for other repairs. If a total of 28 buses were in for maintenance and repairs, how many buses did District A have altogether?
(a) 80 buses
(b) 84 buses
(c) 91 buses
(d) 96 buses

266. On Monday, a kindergarden class uses $2\frac{1}{4}$ kgs of modeling clay the first hour, $4\frac{5}{8}$ kgs of modeling clay the second hour and $\frac{1}{2}$ kg of modeling clay the third hour. How many kgs of clay does the class use during the three hours on Monday?
(a) $6\frac{3}{8}$ kgs
(b) $6\frac{7}{8}$ kgs
(c) $7\frac{1}{4}$ kgs
(d) $7\frac{3}{8}$ kgs

267. Three kittens weigh $2\frac{1}{3}$ kgs, $1\frac{5}{6}$ kgs and $2\frac{2}{3}$ kgs. What is the total weight of the kittens?
(a) $6\frac{1}{3}$ kgs
(b) $6\frac{5}{6}$ kgs
(c) $7\frac{1}{6}$ kgs
(d) $7\frac{1}{3}$ kgs

268. If Ravi has worked a total of $26\frac{1}{4}$ hours so far this week, and has to work a total of $37\frac{1}{2}$ hours, how much longer does he has to work?
(a) $10\frac{1}{4}$ hours
(b) $11\frac{1}{4}$ hours
(c) $11\frac{3}{4}$ hours
(d) $13\frac{1}{2}$ hours

1000 MATH PROBLEMS >>> Fractions

269. On Ravi's daily jog, he travels a distance of $\frac{1}{2}$ mile to get to the track and $\frac{1}{2}$ mile to get home from the track. One lap around the track is $\frac{1}{4}$ mile. If Ravi jogs 5 laps around the track, what is the total distance that he travels?

(a) $2\frac{1}{4}$ miles
(b) $2\frac{1}{2}$ miles
(c) 3 miles
(d) $3\frac{1}{4}$ miles

270. Seema's pie recipe calls for $1\frac{1}{3}$ cups sugar. If she wants to add an additional $\frac{1}{3}$ cup to make the pie sweeter, how much sugar will she need in all?

(a) $1\frac{1}{9}$ cups
(b) $1\frac{1}{6}$ cups
(c) $1\frac{2}{9}$ cups
(d) $1\frac{2}{3}$ cups

271. Amit hiked $7\frac{3}{8}$ miles on Friday, $6\frac{3}{10}$ miles on Saturday, and $5\frac{1}{5}$ miles on Sunday. How many miles did he hike in all?

(a) $18\frac{5}{8}$ miles
(b) $18\frac{7}{8}$ miles
(c) $19\frac{3}{5}$ miles
(d) $20\frac{1}{10}$ miles

272. Mukul owns $16\frac{3}{4}$ acres of land. If he buys another $2\frac{3}{5}$ acres, how many acres of land will he own in all?

(a) $18\frac{4}{5}$ acres
(b) $18\frac{9}{20}$ acres
(c) $19\frac{6}{20}$ acres
(d) $19\frac{7}{20}$ acres

SET 18

273. It takes Mahima 25 minutes to wash her car. If she has been washing her car for 15 minutes, what fraction of the job has she already completed?
    (a) $\frac{3}{5}$
    (b) $\frac{1}{2}$
    (c) $\frac{4}{15}$
    (d) $\frac{2}{5}$

274. It takes 3 firefighters $1\frac{2}{5}$ hours to clean their truck. At that same rate, how many hours would it take one firefighter to clean the same truck?
    (a) $2\frac{4}{7}$ hours
    (b) $3\frac{4}{5}$ hours
    (c) $4\frac{1}{5}$ hours
    (d) $4\frac{2}{5}$ hours

275. How many $5\frac{1}{4}$ ounce glasses can be completely filled from a $33\frac{1}{2}$ ounce container of juice?
    (a) 4 glasses
    (b) 5 glasses
    (c) 6 glasses
    (d) 7 glasses

276. If one pint is $\frac{1}{8}$ of a gallon, how many pints are there in $3\frac{1}{2}$ gallons of ice cream?
    (a) $\frac{7}{16}$ pints
    (b) $24\frac{1}{2}$ pints
    (c) $26\frac{1}{16}$ pints
    (d) 28 pints

277. Rohit's walking speed is $2\frac{1}{2}$ miles per hour. If it takes Rohit 6 minutes to walk from his home to the bus stop, how far is the bus stop from his home?
    (a) $\frac{1}{8}$ mile
    (b) $\frac{1}{4}$ mile
    (c) $\frac{1}{2}$ mile
    (d) 1 mile

278. The directions on an exam allow $2\frac{1}{2}$ hours to answer 50 questions. If you want to spend an equal amount of time on each of the 50 questions, about how much time should you allow for each one?
    (a) 45 seconds
    (b) $1\frac{1}{2}$ minutes
    (c) 2 minutes
    (d) 3 minutes

279. Which of these is equivalent to 35°C? (F = $\frac{9}{5}$C + 32)
    (a) 105°F
    (b) 95°F
    (c) 63°F
    (d) 19°F

1000 MATH PROBLEMS >>> Fractions

280. A firefighter checks the gauge on a cylinder that normally contains 45 cubic feet of air and finds that the cylinder has only 10 cubic feet of air. The guage indicates that the cylinder is

(a) $\frac{1}{4}$ full

(b) $\frac{2}{9}$ full

(c) $\frac{1}{3}$ full

(d) $\frac{4}{5}$ full

281. If the diameter of a metal spool is 3.5 feet, how many times will a 53 foot hose wrap completely around it? $C = \pi d$; $\pi = \frac{22}{7}$

(a) 2 times
(b) 3 times
(c) 4 times
(d) 5 times

282. A person can be scaled by hot water at a temperature of about 122°F. At about what temperature Centigrade could a person be scaled? $C = \frac{5}{9}(F - 32)$

(a) 35.5°C
(b) 55°C
(c) 50°C
(d) 216°C

283. Tank A, when full, holds 555 gallons of water. Tank B, when full, holds 680 gallons of water. If Tank A is only $\frac{2}{3}$ full and Tank B is only $\frac{2}{5}$ full, how many more gallons of water is needed to fill both tanks to capacity?

(a) 319 gallons
(b) 593 gallons
(c) 642 gallons
(d) 658 gallons

284. At a party there are 3 large pizzas. Each pizza has been cut into 9 equal pieces. Eight-ninths of the first pizza has been eaten; $\frac{2}{3}$ of the second pizza has been eaten; $\frac{7}{9}$ of the third pizza has been eaten. What fraction of the 3 pizzas is left?

(a) $\frac{2}{9}$

(b) $\frac{2}{7}$

(c) $\frac{1}{3}$

(d) $\frac{1}{6}$

285. If it takes four firefighters 1 hour and 45 minutes to perform a particular job, how long would it take one firefighter working at the same rate to perform the same task alone?

(a) $4\frac{1}{2}$ hours
(b) 5 hours
(c) 7 hours
(d) $7\frac{1}{2}$ hours

1000 MATH PROBLEMS >>> Fractions

286. If Kiran can change a light bulb in $\frac{5}{6}$ minute, how many minutes would it take him to change five light bulbs?

(a) $4\frac{1}{6}$ minutes

(b) $4\frac{1}{3}$ minutes

(c) $4\frac{2}{3}$ minutes

(d) $5\frac{1}{6}$ minutes

287. Ravi has two bags of jelly beans. One weighs $10\frac{1}{4}$ ounces; the other weighs $9\frac{1}{8}$ ounces. If Ravi puts the two bags together and then divides all of the jelly beans into 5 equal parts to give to his friends, how many ounces will each friend get?

(a) $3\frac{3}{4}$ ounces

(b) $3\frac{7}{8}$ ounces

(c) 4 ounces

(d) $4\frac{1}{4}$ ounces

288. At a certain school, half the students are female and one-twelfth of the students are from outside the state. What proportion of the students would you expect to be females from outside the state?

(a) $\frac{1}{12}$

(b) $\frac{1}{24}$

(c) $\frac{1}{6}$

(d) $\frac{1}{3}$

SET 19

289. How many minutes are in $7\frac{1}{6}$ hours?

(a) 258 minutes
(b) 430 minutes
(c) 2,580 minutes
(d) 4,300 minutes

290. One lap on a particular outdoor track measures a quarter of a mile around. To run a total of three and a half miles, how many complete laps must a person run?

(a) 14 laps
(b) 18 laps
(c) 7 laps
(d) 10 laps

291. During the month of June, Bus #B-461 used the following amounts of oil:

June 1—$3\frac{1}{2}$ quarts

June 19—$2\frac{3}{4}$ quarts

June 30—4 quarts

What is the total number of quarts used in June?

(a) $9\frac{3}{4}$ quarts
(b) 10 quarts
(c) $10\frac{1}{4}$ quarts
(d) $10\frac{1}{2}$ quarts

292. How many ounces are in $9\frac{1}{2}$ pounds?

(a) 192 ounces
(b) 182 ounces
(c) 152 ounces
(d) 132 ounces

293. Mrs. Chawla's yearly income is Rs. 25,000, and the cost of her rent for the year is Rs. 7,500. What fraction of her yearly income does she spend on rent?

(a) $\frac{1}{4}$
(b) $\frac{3}{10}$
(c) $\frac{2}{5}$
(d) $\frac{2}{7}$

294. Mona counts the cars passing her house, and finds that 2 of every 5 cars are foreign. If she counts for an hour, and 60 cars pass, how many of them are likely to be domestic?

(a) 12 cars
(b) 24 cars
(c) 30 cars
(d) 36 cars

295. A recipe calls for $1\frac{1}{4}$ cups of flour. If Larry wants to make $2\frac{1}{2}$ times the recipe, how much flour does he need?

(a) $2\frac{3}{4}$ cups
(b) $3\frac{1}{8}$ cups
(c) $3\frac{1}{4}$ cups
(d) $3\frac{5}{8}$ cups

1000 MATH PROBLEMS >>> Fractions

296. A laboratory technician checks a test tube that normally contains 45 cc of fluid and finds that the test tube has only 10 cc of fluid in it. Her observation indicates that the test tube is

(a) $\frac{1}{4}$ full

(b) $\frac{2}{9}$ full

(c) $\frac{1}{3}$ full

(d) $\frac{4}{5}$ full

297. Third grade student Seema goes to the school nurse's office, where her temperature is found to be 98 degrees Fahrenheit. What is her temperature in degrees Celsius? $C = \frac{5}{9}(F - 32)$

(a) 35.8 degrees C
(b) 36.7 degrees C
(c) 37.6 degrees C
(d) 31.1 degrees C

298. A child has a temperature of 40 degrees C. What is the child's temperature in degrees Fahrenheit? $F = \frac{9}{5}C + 32$.

(a) 100 degrees F
(b) 101 degrees F
(c) 102 degrees F
(d) 104 degrees F

299. If Teena donates Rs. 210.00 to charitable organizations each year and $\frac{1}{3}$ of that amount goes to the local crisis center, how much of her yearly donation does the crisis center get?

(a) Rs. 33.00
(b) Rs. 45.50
(c) Rs. 60.33
(d) Rs. 70.00

300. A construction job calls for $2\frac{5}{6}$ tons of sand. Four trucks, each filled with $\frac{3}{4}$ tons of sand, arrive on the job. Is there enough sand, or is there too much sand for the job?

(a) There is not enough sand; $\frac{1}{6}$ ton more is needed.

(b) There is not enough sand; $\frac{1}{3}$ ton more is needed.

(c) There is $\frac{1}{3}$ ton more sand than is needed.

(d) There is $\frac{1}{6}$ ton more sand than is needed.

1000 MATH PROBLEMS >>> Fractions

301. A safety box has three layers of metal, each with a different width. If one layer is $\frac{1}{8}$ inch thick, a second layer is $\frac{1}{6}$ inch thick, and the total thickness is $\frac{3}{4}$ inch thick, what is the width of the third layer?
 (a) $\frac{5}{12}$ inch
 (b) $\frac{11}{24}$ inch
 (c) $\frac{7}{18}$ inch
 (d) $\frac{1}{2}$ inch

Answer questions 302 and 303 using the following list of ingredients needed to make 16 brownies.

Deluxe Brownies

$\frac{2}{3}$ cup butter

5 squares (1 ounce each) unsweetened chocolate

$1\frac{1}{2}$ cups sugar

2 teaspoons vanilla

2 eggs

1 cup flour

302. What is the greatest number of brownies that can be made if the baker has only 1 cup of butter?
 (a) 12 brownies
 (b) 16 brownies
 (c) 24 brownies
 (d) 28 brownies

303. How much sugar is needed to make 8 brownies?
 (a) $\frac{3}{4}$ cup
 (b) 3 cups
 (c) $\frac{2}{3}$ cup
 (d) $\frac{5}{8}$ cup

304. Girish cuts his birthday cake into 10 equal pieces. If 6 people eat a piece of Girish's cake, what fraction of the cake is left?
 (a) $\frac{3}{5}$
 (b) $\frac{3}{10}$
 (c) $\frac{2}{5}$
 (d) $\frac{5}{6}$

1000 MATH PROBLEMS >>> Fractions

SET 20

305. A certain congressional district has about 490,000 people living in it. The largest city in the area has 98,000 citizens. Which most accurately portrays the portion of the population made up by the city in the district?

(a) $\frac{1}{5}$

(b) $\frac{1}{4}$

(c) $\frac{2}{9}$

(d) $\frac{3}{4}$

306. A bag of jellybeans contains 8 black beans, 10 green beans, 3 yellow beans, and 9 orange beans. What is the probability of selecting either a yellow or an orange bean?

(a) $\frac{1}{10}$

(b) $\frac{2}{5}$

(c) $\frac{4}{15}$

(d) $\frac{3}{10}$

307. Each piece of straight track for Ty's electric train set is $6\frac{1}{2}$ inches long. If 5 pieces of this track are laid end to end, how long will the track be?

(a) $30\frac{1}{2}$ inches

(b) 32 inches

(c) $32\frac{1}{2}$ inches

(d) $32\frac{5}{8}$ inches

308. How many $\frac{1}{4}$ kg hamburgers can be made from 6 kgs of ground beef?

(a) 18 hamburgers

(b) $20\frac{1}{2}$ hamburgers

(c) 24 hamburgers

(d) $26\frac{1}{4}$ hamburgers

309. Bharat's 8-ounce glass is $\frac{4}{5}$ full of water. How many ounces of water does he has?

(a) $4\frac{5}{8}$ ounces

(b) 5 ounces

(c) 6 ounces

(d) $6\frac{2}{5}$ ounces

310. Asha can walk $3\frac{1}{4}$ miles in one hour. At that rate, how many miles will she walk in $1\frac{2}{3}$ hours?

(a) $4\frac{5}{8}$ miles

(b) $4\frac{11}{12}$ miles

(c) $5\frac{5}{12}$ miles

(d) 6 miles

311. Three friends evenly split $1\frac{1}{8}$ kgs of peanuts. How many kgs will each person get?

(a) $\frac{1}{4}$ kg

(b) $\frac{3}{8}$ kg

(c) $\frac{1}{2}$ kg

(d) $\frac{5}{8}$ kg

1000 MATH PROBLEMS >>> Fractions

312. Arun lives $2\frac{1}{2}$ miles due east of the Sunnydale Mall and Rahul lives $4\frac{1}{2}$ miles due west of the QuikMart, which is $1\frac{1}{2}$ miles due west of the Sunnydale Mall. How far does Arun live from Rahul?

(a) 8 miles
(b) 8.5 miles
(c) 9 miles
(d) 9.5 miles

313. Roma's kitchen is $9\frac{3}{4}$ feet long and $8\frac{1}{3}$ feet wide. How many square feet of tile does she need to tile the floor?

(a) $81\frac{1}{4}$ square feet
(b) $72\frac{1}{4}$ square feet
(c) $71\frac{1}{2}$ square feet
(d) $82\frac{1}{2}$ square feet

314. How many inches are in $4\frac{1}{2}$ feet?

(a) 48 inches
(b) 54 inches
(c) 66 inches
(d) 70 inches

315. During the winter, Leela missed $7\frac{1}{2}$ days of kindergarden due to colds, while Bindu missed only $4\frac{1}{4}$ days. How many fewer days did Bindu miss than Leela?

(a) $3\frac{1}{4}$ days
(b) $3\frac{1}{2}$ days
(c) $3\frac{5}{6}$ days
(d) $3\frac{3}{3}$ days

316. To reach his tree house, Raja has to climb $9\frac{1}{3}$ feet up a rope ladder, then $8\frac{5}{6}$ feet up the tree-trunk. How far does Raja has to climb altogether?

(a) $17\frac{7}{12}$ feet
(b) $17\frac{1}{6}$ feet
(c) $18\frac{1}{6}$ feet
(d) $18\frac{1}{2}$ feet

317. Rani's newborn triplets weigh $4\frac{3}{8}$ kgs, $3\frac{5}{6}$ kgs and $4\frac{7}{8}$ kgs. Reema's new born twins weigh $7\frac{2}{6}$ kgs and $9\frac{3}{10}$ kgs. Whose babies weigh the most and by how much?

(a) Rani's triplets by $3\frac{1}{2}$ kgs
(b) Rani's triplets by $2\frac{15}{60}$ kgs
(c) Reema's twins by $1\frac{2}{3}$ kgs
(d) Reema's twins by $3\frac{33}{60}$ kgs.

1000 MATH PROBLEMS >>> Fractions

318. Mohan's mother gave him a pound of fudge. After eating $\frac{5}{8}$ of his fudge, Mohan stopped because he suddenly felt queasy. How many ounces of fudge had Mohan eaten when he became ill?
   - (a) 9 ounces
   - (b) 10 ounces
   - (c) 11 ounces
   - (d) 12 ounces

319. Beena received a surprise gift from her grandmother of Rs. 4,500. She immediately spent Rs 450 on a fake rabbit-fur coat. What fraction of her gift did she spend on the coat?
   - (a) $\frac{1}{10}$
   - (b) $\frac{1}{18}$
   - (c) $\frac{1}{20}$
   - (d) $\frac{1}{40}$

320. During an 8 hour work-day, Bunty spends 2 hours on the phone. What fraction of the day does he spend on the phone?
   - (a) $\frac{2}{10}$
   - (b) $\frac{1}{3}$
   - (c) $\frac{1}{4}$
   - (d) $\frac{1}{8}$

SECTION

3

DECIMALS

These 10 sets of problems will familiarize you with arithmetic operations involving decimals (which are a really special kind of fraction). You use decimals every day, in dealing with money, for example. Units of measurements, such as populations, kilometers, inches, or miles are also often expressed in decimals. In this section you will get practice in working with *mixed decimals*, or numbers that have digits on both sides of a decimal point, and with the important tool of *rounding*, the method for estimating decimals.

1000 MATH PROBLEMS >>> Decimals

SET 21

321. 56.73647 rounded to the nearest hundredth is equal to
    (a) 100
    (b) 57
    (c) 56.7
    (d) 56.74

322. Which number sentence is true?
    (a) 0.43 < 0.043
    (b) 0.0043 > 0.43
    (c) 0.00043 > 0.043
    (d) 0.043 > 0.0043

323. 78.09 + 19.367 =
    (a) 58.723
    (b) 87.357
    (c) 97.457
    (d) 271.76

324. 3.419 − 0.7 =
    (a) 34.12
    (b) 0.2719
    (c) 2.719
    (d) 0.3412

325. 2.9 ÷ 0.8758
    (a) 3.31
    (b) 0.331
    (c) 0.302
    (d) 0.0302

326. 195.6 ÷ 7.2, rounded to the nearest hundredth, is equal to
    (a) 271.67
    (b) 27.17
    (c) 27.16
    (d) 2.717

327. 426 − 7.2 =
    (a) 354.0
    (b) 425.28
    (c) 418.8
    (d) 41.88

328. 6.35 × 5 =
    (a) 31.75
    (b) 30.75
    (c) 3.175
    (d) 317.5

329. 5.9 − 4.166 =
    (a) 1.844
    (b) 1.843
    (c) 1.744
    (d) 1.734

330. 172 × 0.56 =
    (a) 9.632
    (b) 96.32
    (c) 963.2
    (d) 0.9632

331. 0.63 × 0.42 =
    (a) 26.46
    (b) 2.646
    (c) 0.2646
    (d) 0.02646

332. What is six and five hundredths written as a decimal?
    (a) 0.065
    (b) 6.5
    (c) 6.05
    (d) 6.005

1000 MATH PROBLEMS >>> Decimals

333. In the following decimal, which digit is in the hundredths place? 0.9402
(a) 9
(b) 0
(c) 2
(d) 4

334. What is 0.716 rounded to the nearest tenth?
(a) 0.7
(b) 0.8
(c) 0.72
(d) 1.00

335. Which of these has a 9 in the thousandths place?
(a) 3.0095
(b) 3.0905
(c) 3.9005
(d) 3.0059

336. 4.5 ÷ 2.5 =
(a) 20.0
(b) 2.0
(c) 1.8
(d) 0.2

SET 22

337. 0.216 + 9.3 + 72.0 =
(a) 81.516
(b) 77.516
(c) 16.716
(d) 0.381

338. 0.49 × 0.07 =
(a) 34.3
(b) 0.0343
(c) 3.43
(d) 0.343

339. 945.6 ÷ 24 =
(a) 3,940
(b) 394
(c) 39.4
(d) 3.946

340. 72.687 + 145.29 =
(a) 87.216
(b) 217.977
(c) 217.877
(d) 882.16

341. 0.088 + 0.091 =
(a) 0.017
(b) 0.169
(c) 0.177
(d) 0.179

342. 3.6 − 1.89 =
(a) 1.47
(b) 1.53
(c) 1.71
(d) 2.42

343. What is another way to write 2.75×100^2?
(a) 275
(b) 2,750
(c) 27,500
(d) 270,000

344. 3.16 ÷ 0.079 =
(a) 0.025
(b) 2.5
(c) 4.0
(d) 40.0

345. 367.08 × 0.15 =
(a) 22.0248
(b) 55.051
(c) 55.062
(d) 540.62

346. What is another way to write 7.25×10^3?
(a) 72.5
(b) 725
(c) 7,250
(d) 72,500

347. $(4.1 \times 10^{-2})(3.8 \times 10^4) =$
(a) 1.558×10^{-8}
(b) 15.58×10^{-2}
(c) 1.558×10^2
(d) 1.558×10^3

348. $\frac{6.5 \times 10^{-6}}{3.25 \times 10^3} =$
(a) 2×10^{-9}
(b) 2×10^{-3}
(c) 2×10^2
(d) 2×10^3

1000 MATH PROBLEMS >>> *Decimals*

349. What is the result of multiplying 11 by 0.032?
 (a) 0.032
 (b) 0.0352
 (c) 0.32
 (d) 0.352

350. 0.31 + 0.673 =
 (a) 0.0983
 (b) 0.983
 (c) 0.967
 (d) 9.83

351. Which of the following numbers is NOT between −0.02 and 1.02?
 (a) −0.15
 (b) −0.015
 (c) 0
 (d) 0.02

SET 23

352. What is the decimal 0.0008 written out in words?
    (a) eight tenths
    (b) eight hundredths
    (c) eight thousandths
    (d) eight ten-thousandths

353. Which of the following decimals has the greatest value?
    (a) 6.723
    (b) 6.0723
    (c) 6.7023
    (d) 6.7

354. Which of the following decimals has the LEAST value?
    (a) 0.0012
    (b) 0.0102
    (c) 0.012
    (d) 0.12

355. What is 72.0094 rounded to the nearest hundredth?
    (a) 72.009
    (b) 72.01
    (c) 72.09
    (d) 72.1

356. What is the sum of 11.006 + 34 + 0.72 rounded to the nearest tenth?
    (a) 45.1
    (b) 45.7
    (c) 45.73
    (d) 46

357. 12.9991 + 78 + 0.9866 =
    (a) 90.9857
    (b) 90.0957
    (c) 91.9857
    (d) 91.9957

358. 2 − 0.42 − 0.1 =
    (a) 14.8
    (b) 1.48
    (c) 1.048
    (d) 1.0048

359. Noorie and Julie begin driving at the same time but in opposite directions. If Noori drives 60 miles per hour, and Julie drives 70 miles per hour, how long will it be before they are 325.75 miles apart?
    (a) 2 hours
    (b) 2.25 hours
    (c) 2.5 hours
    (d) 2.75 hours

360. A supermarket offers 25% off on all pumpkins one day, just before Halloween. If a certain pumpkin is sale-priced at Rs 4.20, what was the original price?
    (a) Rs. 5.62
    (b) Rs. 5.60
    (c) Rs. 5.58
    (d) Rs. 5.56

361. A bottle of merlot is 1.5 litres; a bottle of chardonnay is 1.35 litres. How many total litres of wine is in the two bottles?
    (a) 2.35 litres
    (b) 2.50 litres
    (c) 2.85 litres
    (d) 2.90 litres

1000 MATH PROBLEMS >>> Decimals

362. Over the last few years, the average number of dogs in the neighbourhood has dropped from 17.8 to 14.33. What is the decrease in the average number of dogs in the neighbourhood?
(a) 3.47 dogs
(b) 2.66 dogs
(c) 3.15 dogs
(d) 2.75 dogs

363. On Monday, Farida slept for 6.45 hours; on Tuesday, 7.32; on Wednesday, 5.1; on Thursday, 6.7; and on Friday she slept for 8.9 hours. How much total sleep did she get over the five days?
(a) 40.34 hours
(b) 34.47 hours
(c) 36.78 hours
(d) 42.67 hours

364. Last week, Rashmi had Rs. 679.80 saved from baby-sitting. She made another Rs 157.50 baby-sitting this week and spent 275.80 on CDs. How much money does she has now?
(a) Rs. 715.50
(b) Rs. 246.50
(c) Rs. 1113.10
(d) Rs. 561.50

365. Roommates Ravi and Raj agreed to wallpaper and carpet the living room and replace the sofa. The wallpaper costs Rs.103.84, the carpet costs Rs. 598.15, and the new sofa costs Rs. 768.56. Ravi agrees to pay for the carpet and wallpaper and Raj agrees to pay for the sofa. How much more money will Raj spend than Ravi?
(a) Rs. 68.56
(b) Rs. 76.55
(c) Rs. 66.57
(d) Rs. 72.19

366. Kitty litter costs Rs. 3.59 for 25 kgs. How much does 100 kgs cost?
(a) Rs. 14.36
(b) Rs. 35.90
(c) Rs. 17.95
(d) Rs. 10.77

367. A late night talk show host got a total of 63 laughs over five nights. How many laughs did he average per night?
(a) 5.6 laughs
(b) 14.8 laughs
(c) 6.3 laughs
(d) 12.6 laughs

368. Six friends agree to evenly split the cost of gasoline on a trip. Each friend paid Rs. 37.27. What was the total cost of gas?
(a) Rs. 370.27
(b) Rs. 223.62
(c) Rs. 314.78
(d) Rs. 262.78

SET 24

369. Sandip paid Rs. 5.96 for 4 kgs of cookies. How much do the cookies cost per kg.
(a) Rs. 1.96 per kg
(b) Rs. 2.33 per kg
(c) Rs. 1.49 per kg
(d) Rs. 2.15 per kg

370. One inch equals 2.54 centimeters. How many centimeters are there in a foot?
(a) 30.48 centimeters
(b) 32.08 centimeters
(c) 31.79 centimeters
(d) 29.15 centimeters

371. Meena and Jona can restock an aisle at the supermarket in one hour working together. Meena can restock an aisle in 1.5 hours working alone, and it takes Jona 2 hours to restock an aisle. If they work together for two hours, and then work separately for another two hours, how many aisles will they have completed?
(a) 5
(b) 4.5
(c) 4.33
(d) 3.5

372. Pran earns only $\frac{1}{8}$ what Hari does. Hari makes Rs. 19.50 an hour. For an 8-hour day, how much does Pran earn?
(a) Rs. 18.50
(b) Rs. 18.75
(c) Rs. 19.50
(d) Rs. 19.75

373. Reema earns Rs. 10 an hour for walking the neighbour's dog. Today she can only walk the dog for 45 minutes. How much will Reema make today?
(a) Rs. 6.25
(b) Rs. 7.50
(c) Rs. 7.75
(d) Rs. 8.00

374. My greyhound, Heera, can run 35.25 miles an hour, while my cat, Spot, can run only $\frac{1}{4}$ that fast. How many miles per hour can Spot run?
(a) 8.25 miles per hour
(b) 8.77 miles per hour
(c) 8.81 miles per hour
(d) 9.11 miles per hour

375. A 600-page book is 1.5 inches thick. What is the thickness of each page?
(a) 0.0010 inches
(b) 0.0030 inches
(c) 0.0025 inches
(d) 0.0600 inches

376. An ant starts climbing straight up a wall at 6:07 a.m. Ten minutes later it reaches the ceiling. If the ant travels steadily at 48.5 feet per hour, how high is the ceiling?
(Distance = rate × time)
(a) 4.8 feet
(b) 8.0 feet
(c) 10.8 feet
(d) 48.1 feet

1000 MATH PROBLEMS >>> Decimals

377. For every rupee Kiran saves, her employer contributes a dime to her savings, with a maximum employer contribution of Rs. 10 per month. If Kiran saves Rs. 60 in January, Rs. 130 in March, and Rs. 70 in April, how much will she has in savings at the end of that time?
(a) Rs 270
(b) Rs 283
(c) Rs 286
(d) Rs 290

378. Giri was shopping for a new washing machine. The one he wanted to buy cost Rs. 428.98. The salesperson informed him that the same machine would be on sale the following week for Rs. 399.99. How much money would Hari save by waiting until the washing machine went on sale?
(a) Rs. 28.99
(b) Rs. 29.01
(c) Rs. 39.09
(d) Rs. 128.99

379. If a bus weighs 2.5 tons, how many pounds does it weigh? (1 ton = 2,000 pounds)
(a) 800
(b) 4,500
(c) 5,000
(d) 5,500

380. If Seema burns about 304.15 calories while walking fast on her treadmill for 38.5 minutes, about how many calories does she burn per minute?
(a) 7.8
(b) 7.09
(c) 7.9
(d) 8.02

381. A truck is carrying 1,000 television sets; each set weighs 21.48 kgs. What is the total weight, in kgs, of the entire load?
(a) 214.8
(b) 2,148
(c) 21,480
(d) 214,800

382. Leena is mailing two packages. One weighs 12.9 kgs and the other weighs half as much. What is the total weight in kgs of the two packages?
(a) 6.45
(b) 12.8
(c) 18.5
(d) 19.35

383. If it takes Geeta 22.4 minutes to walk 1.25 miles, how many minutes will it take her to walk one mile?
(a) 17.92
(b) 18
(c) 19.9
(d) 21.15

384. Manish's temperature at 9:00 a.m. was 97.2° F. At 4:00 p.m., his temperature was 99° F. By how many degrees did his temperature rise?
(a) 0.8
(b) 1.8
(c) 2.2
(d) 2.8

1000 MATH PROBLEMS >>> Decimals

SET 25

385. Jeny had Rs. 40.00 in his wallet. He bought gasoline for Rs. 12.90, a pack of gum for Rs. 0.45, and a candy bar for Rs. 0.88. How much money did he has left?
    (a) Rs. 14.23
    (b) Rs. 25.77
    (c) Rs. 25.67
    (d) Rs. 26.77

386. The price of cheddar cheese is Rs. 2.12 per kg. The price of Monterey Jack cheese is Rs. 2.34 per kg. If Hari buys 1.5 kgs of cheddar and 1 kg of Monterey Jack, how much will he spend in all?
    (a) Rs. 3.18
    (b) Rs. 4.46
    (c) Rs. 5.41
    (d) Rs. 5.52

387. From a 100-foot ball of string, Romy cuts three pieces of the following lengths: 5.8 feet, 3.2 feet, 4.4 feet. How many feet of string are left?
    (a) 66
    (b) 86.6
    (c) 87.6
    (d) 97.6

388. Fifteen-ounce cans of clam chowder sell at 3 for Rs. 2.00. How much does one can cost, rounded to the nearest paise?
    (a) Rs. 0.60
    (b) Rs. 0.66
    (c) Rs. 0.67
    (d) Rs. 0.70

389. For the month of July, Veena purchased the following amounts of gasoline for her car: 9.4 gallons, 18.9 gallons, and 22.7 gallons. How many gallons did she purchase in all?
    (a) 51
    (b) 51.9
    (c) 52
    (d) 61

390. Manish works Monday through Friday each week. His bus fare to and from work is Rs. 1.10 each way. How much does Manish spend on bus fare each week?
    (a) Rs. 10.10
    (b) Rs. 11.00
    (c) Rs. 11.10
    (d) Rs. 11.20

391. If one centimetre equals 0.39 inches, how many centimetres are there in 0.25 inches?
    (a) 1.56
    (b) 0.641
    (c) 0.0273
    (d) 0.0975

392. 5.133 multiplied by 10^{-6} is equal to
    (a) 0.0005133
    (b) 0.00005133
    (c) 0.000005133
    (d) 0.0000005133

393. On a business trip, Meera went out to lunch. The shrimp cocktail cost Rs. 5.95, the blackened swordfish with grilled vegetables cost Rs. 11.70, the cherry cheesecake cost Rs. 4.79, and the coffee was Rs. 1.52. What was Meera's total bill?
    (a) Rs. 23.96
    (b) Rs. 26.93
    (c) Rs. 29.63
    (d) Rs. 32.99

1000 MATH PROBLEMS >>> Decimals

394. Hari wants to buy a used car to take to college. The car costs Rs. 4999.95. For graduation, he receives gifts of Rs. 200.00, Rs. 157.75 and Rs. 80.50. His little brother gave him Rs. 1.73 and he saved Rs. 4332.58 from his summer job. How much more money does he need?

(a) Rs. 272.93
(b) Rs. 227.39
(c) Rs. 722.93
(d) Rs. 772.39

395. A football team must move the ball ten yards in four plays in order to keep possession of the ball. The fighting Jellyfish have just run a play in which they gained 2.75 yards. How many yards remain?

(a) 6.50 yards
(b) 6.75 yards
(c) 7.25 yards
(d) 7.50 yards

396. Harish wants to build a deck that is 8.245 feet by 9.2 feet. How many square feet will the deck be? (Area = length × width).

(a) 34.9280 square feet
(b) 82.4520 square feet
(c) 17.4450 square feet
(d) 75.8540 square feet

397. Ravi's car used 23.92 gallons of fuel on a 517.6 mile trip. How many miles per gallon did the car get, rounded to the nearest hundredth?

(a) 18.46 gallons
(b) 21.64 gallons
(c) 26.14 gallons
(d) 29.61 gallons

398. Farida ran a 200-meter race in 23.7 seconds. Mansi ran it in 22.59 seconds. How many fewer seconds did it take Mansi than Farida?

(a) 2.17 seconds
(b) 0.97 seconds
(c) 1.01 seconds
(d) 1.11 seconds

399. Mona buys 2.756 kgs of sliced turkey, 3.2 kgs of roast beef, and 5.59 kgs of bologna. Approximately how many total kilos of meat did Mona buy?

(a) 11.5 kgs
(b) 12.75 kgs
(c) 10.03 kgs
(d) 13.4 kgs

400. Sridhar reads 25.56 pages of his law textbook in an hour. How many pages can he read in 5 hours?

(a) 51.58 pages
(b) 76.75 pages
(c) 127.80 pages
(d) 142.96 pages

1000 MATH PROBLEMS >>> Decimals

SET 26

401. Indra has 7.5 kgs of candy for trick-or-treaters. She gives a vampire 0.25 kgs, a fairy princess 0.53 kgs, and a horse 1.16 kgs. She eats 0.12 kgs. How much candy is left?

(a) 5.44 kgs
(b) 4.55 kgs
(c) 5.3 kgs
(d) 4.7 kgs

402. Meena drives from home to the grocery, which is 6.2 miles. Then she goes to the video store, which is 3.4 more miles. Next, she goes to the bakery, which is 0.82 more miles. Then she drove the 5.9 miles home. How many miles total did she drive?

(a) 12.91 miles
(b) 13.6 miles
(c) 16.32 miles
(d) 18.7 miles

403. Kiran wants to make slip covers for his dining room chairs. Each chair requires 4.75 metres of fabric. Kiran has 20.34 metres of fabric. How many chairs can he cover?

(a) 3
(b) 4
(c) 5
(d) 6

404. On Monday, Leela had Rs. 792.78 in her checking account. On Tuesday, she deposited her Rs. 1252.60 paycheck. On Wednesday she paid her rent, Rs. 650. On Thursday, she pair her electric, cable, and phone bills, which were Rs. 79.35, Rs. 54.23, Rs. 109.56 respectively. How much money is left in Leela's account?

(a) Rs. 965.73
(b) Rs. 1348.90
(c) Rs. 893.14
(d) Rs. 1152.24

405. Meera has a frame that is 3.34 feet by 2.685 feet. She needs to buy a canvas for the frame. What is the area, rounded to the nearest hundredth, of the canvas she should buy?

(a) 8.97 square feet
(b) 6.02 square feet
(c) 7.52 square feet
(d) 5.92 square feet

406. Manish's favourite cake recipe calls for 0.75 kgs of flour; he has a 5 kg bag. He wants to make several cakes for the school bake sale. How many cakes can he make?

(a) 5
(b) 6
(c) 7
(d) 8

1000 MATH PROBLEMS >>> Decimals

407. Harini walks to work every day, sometimes running errands on the way. On Monday she walked 0.75 mile; Tuesday she walked 1.2 miles; Wednesday she walked 1.68 miles; Thursday she walked 0.75 mile; Friday she rode with Earl. On the days she walked, what was the average distance Harini walked each day?

(a) 2.78 miles
(b) 1.231 miles
(c) 0.75 miles
(d) 1.095 miles

408. Seven people are at the beach for a clambake. They have dug 12.6 kgs of clams. They each eat the following amounts of clams: 0.34 kgs, 1.6 kgs, 0.7 kgs, 1.265 kgs, 0.83 kgs, 1.43 kgs, 0.49 kgs. How many kilos of clams are left?

(a) 7.892 kgs
(b) 4.56 kgs
(c) 5.945 kgs
(d) 6.655 kgs

409. A famous actor will be paid 13.8 million dollars for his next film. His co-star, a famous actress, will be paid 4.892 million. How much less is the actress making than the actor?

(a) 8.908 million
(b) 7.542 million
(c) 6.89 million
(d) 5.982 million

410. A sisal carpet costs Rs. 3.78 per square foot. How much carpet, rounded to the nearest square foot, could Jerry buy with Rs. 75?

(a) 12 square feet
(b) 15 square feet
(c) 20 square feet
(d) 23 square feet

411. Pradeep has worked 34.75 hours of his usual 39.5 hour week. How many hours does he has left to work?

(a) 5.25 hours
(b) 4.75 hours
(c) 4.00 hours
(d) 3.75 hours

412. Roma has to plough a 0.75 acre field Monday, a 1.4 acres on field on Tuesday, and a 0.68 acre field on Wednesday. How many acres does she has to plough in the three days?

(a) 3.07 acres
(b) 2.15 acres
(c) 3.92 acres
(d) 2.83 acres

413. Lalita is going to re-carpet her living room. The dimensions of the room are 15.6 feet by 27.75 feet. How many square feet of carpet will she need?

(a) 315.8 square feet
(b) 409.5 square feet
(c) 329.25 square feet
(d) 432.9 square feet

1000 MATH PROBLEMS >>> Decimals

414. A movie is scheduled for two hours. The theatre advertisements are 3.8 minutes long. There are two previews; one is 4.6 minutes long and the other is 2.9 minutes. The rest of the time is devoted to the feature. How long is the feature film?

(a) 108.7 minutes
(b) 97.5 minutes
(c) 118.98 minutes
(d) 94.321 minutes

415. Tony is making hats for the kids in the neighbourhood. He needs 0.65 metre of fabric for each hat; he has 10 metres of fabric. How many hats can he make?

(a) 13
(b) 14
(c) 15
(d) 16

416. Over the weekend, Mahima watched 11.78 hours of television, Jona watched 6.9 hours, Chetna watched 7 hours, and Mona watched 2.45 hours. How much television did they watch over the weekend?

(a) 26.786 hours
(b) 28.13 hours
(c) 30.79 hours
(d) 32.85 hours

1000 MATH PROBLEMS >>> Decimals

SET 27

417. Mona wants to fence in her backyard for her dog. The yard is 78.45 feet by 65.89 feet. How many feet of fence will she need?

(a) 288.68 feet
(b) 144.34 feet
(c) 245.89 feet
(d) 167.17 feet

418. In the new television season, an average of 7.9 million people watched one network; an average of 8.6 million people watched another. How many more viewers did the second network average than the first?

(a) 0.5 million
(b) 0.6 million
(c) 0.7 million
(d) 0.8 million

419. A writer makes Rs. 1.13 per book sold. How much will she make when 100 books have been sold?

(a) Rs. 11.30
(b) Rs. 113.00
(c) Rs. 1,130.00
(d) Rs. 11,300.00

420. Mani walks 0.75 mile to school; Roma walks 1.3 miles; Rama walks 2.8 miles; and Prasana walks 0.54 miles. What is the total distance the four walk to school?

(a) 4.13 miles
(b) 5.63 miles
(c) 4.78 miles
(d) 5.39 miles

421. Rashmi is driving 46.75 miles per hour. How far will she go in 15 minutes?

(a) 14.78 miles
(b) 11.6875 miles
(c) 12.543 miles
(d) 10.7865 miles

422. After stopping at a rest stop, Rashmi continues to drive at 46.75 miles per hour. How far will she go in 3.80 hours?

(a) 177.65 miles
(b) 213.46 miles
(c) 143.78 miles
(d) 222.98 miles

423. While doing her family tree, Beena studies 88 or her ancestors and learns that, in 1673, the ratio of pirates to clergy was 3:4. How many of her relatives were pirates in 1673?

(a) 33
(b) 44
(c) 55
(d) 66

424. District C spends about Rs. 4,446.00 on diesel fuel each week. If the cost of diesel fuel to the district is about Rs. 11.70 per litre about how many litres of diesel fuel does the district use in one week?

(a) 380 litres
(b) 381 litres
(c) 397.2 litres
(d) 520.2 litres

1000 MATH PROBLEMS >>> Decimals

425. If one inch equals 2.54 centimetres, how many inches are there in 20.32 centimetres?
(a) 7.2
(b) 8
(c) 9
(d) 10.2

426. Kiran earns Rs. 12.50 an hour. When she works more than 8 hours in one day, she earns $1\frac{1}{2}$ times her regular hourly wage. If she earns Rs. 137.50 for one day's work, how many hours did she work that day?
(a) 8.5
(b) 9
(c) 10
(d) 11

427. If a state police car travels at the speed of 62 mph for 15 minutes, how far will it travel?
(Distance = rate × time)
(a) 9.3 miles
(b) 15.5 miles
(c) 16 miles
(d) 24.8 miles

428. If the speed of light is 3.00×10^8 metres per second, how far would a beam of light travel in 2,000 seconds?
(a) 1.50×10^5 metres
(b) 6.00×10^5 metres
(c) 1.50×10^{11} metres
(d) 6.00×10^{11} metres

429. Which of the following rope lengths is the longest? (1 cm = 0.39 inches)
(a) 1 metre
(b) 1 yard
(c) 32 inches
(d) 85 centimetres

430. If a law enforcement officer weighs 168 pounds, what is the approximate weight of that officer in kilograms? (1 kilogram = about 2.2 pounds)
(a) 76
(b) 77
(c) 149
(d) 150

431. If a worker is given a pay increase of Rs. 1.25 per hour, what is the total amount of the pay increase for one 40-hour week?
(a) Rs. 49.20
(b) Rs. 50.00
(c) Rs. 50.25
(d) Rs. 51.75

432. The cost of a list of supplies for a fire station is as follows: Rs. 19.98, Rs. 52.20, Rs. 12.64 and Rs. 7.79. What is the total cost?
(a) Rs. 91.30
(b) Rs. 92.61
(c) Rs. 93.60
(d) Rs. 93.61

SET 28

433. A firefighter determines that the length of hose needed to reach a particular building is 175 feet. If the available hoses are 45 feet long, how many sections of hose, when connected together, will it take to reach the building?
(a) 2
(b) 3
(c) 4
(d) 5

434. About how many litres of water will a 5-gallon container hold? (1 litre = 1.06 quarts)
(a) 5
(b) 11
(c) 19
(d) 21

435. If one gallon of water weighs 8.35 kgs, a 25-gallon container of water would most nearly weigh
(a) 173 kgs
(b) 200 kgs
(c) 209 kgs
(d) 215 kgs

436. Raja wants to know if he has enough money to purchase several items. He needs three heads of lettuce, which cost Rs. 0.99 each, and two boxes of cereal, which cost Rs. 3.49 each. He uses the expression (3 × Rs. 0.99) + (2 × Rs. 3.49) to calculate how much the items will cost. Which of the following expressions could also be used?
(a) 3 × (Rs. 3.49 + Rs. 0.99) − Rs 3.49
(b) 3 × (Rs. 3.49 + Rs. 0.99)
(c) (2 + 3) × (Rs. 3.49 + Rs. 0.99)
(d) (2 × 3) + (Rs. 3.49 × Rs. 0.99)

437. If you take recyclables to whichever recycler will pay the most, what is the greatest amount of money you could get for 2,200 kgs of aluminium, 1,400 kgs of cardboard, 3,100 kgs of glass, and 900 kgs of plastic?

Recycler	Aluminium	Cardboard	Glass	Plastic
X	.06/kg	.03/kg	.08/kg	.02/kg
Y	.07/kg	.04/kg	.07/kg	.03/kg

(a) Rs. 409
(b) Rs. 440
(c) Rs. 447
(d) Rs. 485

438. If the average person throws away 3.5 kgs of trash every day, how much trash would the average person throw away in one week?
(a) 24.5 kgs
(b) 31.5 kgs
(c) 40.2 kgs
(d) 240 kgs

439. In production line A can produce 12.5 units in an hour, and production line B can produce 15.25 units in an hour, how long will production line A have to work to produce the same amount of units as line B?
(a) 1 hour
(b) 1.22 hours
(c) 1.50 hours
(d) 1.72 hours

440. Raj earns Rs. 12.50 for each hour that he works. If Raj works 8.5 hours per day, five days a week, how much does he earn in a week?
(a) Rs. 100.00
(b) Rs. 106.25
(c) Rs. 406.00
(d) Rs. 531.25

1000 MATH PROBLEMS >>> Decimals

441. Shimla recently received a snow storm that left a total of eight inches of snow. If it snowed at a consistent rate of three inches every two hours, how much snow had fallen in the first five hours of the storm?
(a) 3 inches
(b) 3.3 inches
(c) 5 inches
(d) 7.5 inches

442. A family eats at Nirula's and orders the following items from the menu:
Hamburger   Rs. 20.95
Dosa   Rs. 30.35
Chicken Sandwich Rs. 30.95
Grilled Cheese Rs. 10.95

If the family orders 2 hamburgers, 1 dosa, 2 chicken sandwiches, and 1 grilled cheese, what is the total cost of their order?
(a) Rs. 151.15
(b) Rs. 171.25
(c) Rs. 180.05
(d) Rs. 145.10

443. If a physical education student burns 8.2 calories per minute while riding a bicycle, how many calories will the same student burn if she rides for 35 minutes?
(a) 246 calories
(b) 286 calories
(c) 287 calories
(d) 387 calories

444. It takes a typing student 0.75 seconds to type one word. At this rate, how many words can the student type in 60 seconds?
(a) 8.0 words
(b) 45.0 words
(c) 75.0 words
(d) 80.0 words

445. Khanna's Market sells milk for Rs. 12.24 per litre. Food Supply sells the same milk for Rs. 12.08 per litre. If Rani buys 2 litres of milk at Food Supply instead of Khanna's, how much will she save?
(a) Rs. 0.12
(b) Rs. 0.14
(c) Rs. 0.32
(d) Rs. 0.38

446. An office uses 2 dozen pencils and $3\frac{1}{2}$ reams of paper each week. If pencils cost 50 paise each and a ream of paper costs Rs. 75, how much does it cost to supply the office for a week?
(a) 75.5
(b) 122.0
(c) 262.5
(d) 274.5

447. If a particular woman's resting heartbeat is 72 beats per minute and she is at rest for $6\frac{1}{2}$ hours, about how many times will her heart beat during that period of time?
(a) 4,320
(b) 4,680
(c) 28,080
(d) 43,200

448. Veena earns 1.5 times less money per hour than Girish does. If Veena earns Rs. 6.50 per hour, how much per hour does Girish earn?
(a) Rs. 8.00
(b) Rs. 8.25
(c) Rs. 9.55
(d) Rs. 9.75

1000 MATH PROBLEMS >>> Decimals

SET 29

449. It takes five-year-old Giri 1.6 minutes to tie the lace on his right shoe and 1.5 minutes to tie the lace on his left shoe. How many minutes does it take Giri to tie both shoes?

(a) 2.1
(b) 3.0
(c) 3.1
(d) 4.1

450. Asha rode her bicycle a total of 25.8 miles in 3 days. On average, how many miles did she ride each day?

(a) 8.06
(b) 8.6
(c) 8.75
(d) 8.9

451. If one inch equals 2.54 centimetres, how many inches are there in 254 centimetres?

(a) $\frac{1}{10}$
(b) 10
(c) 100
(d) 1000

452. Prem ran 6.45 miles on Monday, 5.9 miles on Tuesday, and 6.75 miles on Wednesday. What is the total number of miles Prem ran?

(a) 19.1
(b) 19.05
(c) 17
(d) 13.79

453. Satish's resting heart rate is about 71 beats per minute. If Satish is at rest for 35.2 minutes, about how many times will his heart beat during that period of time?

(a) 2398.4
(b) 2408.4
(c) 2490.3
(d) 2499.2

454. If one kg of chicken costs 82.75, how much does 0.89 kg of chicken cost, rounded to the nearest paise?

(a) Rs. 2.40
(b) Rs. 2.48
(c) Rs. 2.68
(d) Rs. 4.72

455. On Wednesday morning, Yogi's Appliance Service had balance of Rs. 2,354.82 in its checking account. If the bookkeeper wrote a total of Rs. 867.59 worth of checks that day, how much was left in the checking account?

(a) Rs. 1487.23
(b) Rs. 1487.33
(c) Rs. 1496.23
(d) Rs. 1587.33

456. If Nita cuts a length of ribbon that is 13.5 inches long into 4 equal pieces, how long will each piece be?

(a) 3.3075
(b) 3.375
(c) 3.385
(d) 3.3805

1000 MATH PROBLEMS >>> Decimals

457. At age Sixteen, Ramesh weighed 40.6 kgs. By age Eighteen, Ramesh weighed 46.1 kgs. How much weight did he gain in that one year?

(a) 4.5 kgs
(b) 5.5 kgs
(c) 5.7 kgs
(d) 6.5 kgs

458. While on a three-day vacation, Varun spent the following amounts on motel rooms: Rs. 52.50, Rs. 47.99, Rs. 49.32. What is the total amount he spent?

(a) Rs. 139.81
(b) Rs. 148.81
(c) Rs. 148.83
(d) Rs. 149.81

459. Mani grew 0.6 inches during his senior year in high school. If he was 68.8 inches tall at the beginning of his senior year, how tall was he at the end of the year?

(a) 69
(b) 69.2
(c) 69.4
(d) 74.8

460. For a science project, Sita and Tina are measuring the length of two caterpillars. Sita's caterpillar is 2.345 centimeters long. Tina's caterpillar is 0.0005 centimeters longer. How long is Tina's caterpillar?

(a) 2.0345
(b) 2.3455
(c) 2.0345
(d) 2.845

461. About how many quarts of water will a 3.25-litre container hold? (1 litre = 1.06 quarts).

(a) 3.066
(b) 3.045
(c) 3.445
(d) 5.2

462. Nishi has basic cable television service at a cost of Rs. 139.50 per month. If she adds the movie channels, it will cost an additional Rs. 57.00 per month. The sports channels cost another Rs. 48.90 per month. If Nishi adds the movie channels and the sports channels, what will her total monthly payment be?

(a) Rs. 235.4
(b) Rs. 235.5
(c) Rs. 245.4
(d) Rs. 345.4

463. The fares collected for one bus on Route G47 on Monday are as follows: Run 1—Rs. 419.50, Run 2—Rs. 537.00, Run 3—Rs. 390.10, Run 4—Rs. 425.50. What is the total amount collected?

(a) Rs. 1,661.10
(b) Rs. 1,762.20
(c) Rs. 1,772.10
(d) Rs. 1,881.00

464. If it takes Bunny and Sunny about 0.67 hours to mow one half-acre lawn, about how many hours would it take Bunny alone to mow 5 half-acre lawns?

(a) 3.35
(b) 4.35
(c) 5.75
(d) 6.7

1000 MATH PROBLEMS >>> Decimals

SET 30

465. The town of Crystal Point collected Rs. 84,493.26 in taxes last year. This year, the town collected Rs. 91,222.30 in taxes. How much more money did the town collect this year?
(a) Rs. 6,729.04
(b) Rs. 6,729.14
(c) Rs. 6,739.14
(d) Rs. 7,829.04

466. It took Dinesh 3.75 hours to drive 232.8 miles. What was his average mile-per-hour speed?
(a) 62.08
(b) 62.8
(c) 63.459
(d) 71.809

467. Meena has budgeted Rs. 1,000/- for the week to spend on food. If she buys Rice that costs Rs. 128.40 and 4 kgs of Dal that costs Rs. 31.60 per kg, how much of her weekly food budget will she have left?
(a) Rs. 745.20
(b) Rs. 800.00
(c) Rs. 840.00
(d) Rs. 866.20

468. Three 15.4-inch pipes are laid end to end. What is the total length of the pipes in feet?
(1 foot = 12 inches)
(a) 3.02
(b) 3.2
(c) 3.85
(d) 4.62

469. If one ounce equals 28.571 grams, 12.1 ounces is equal to how many grams?
(a) 37.63463
(b) 343.5473
(c) 345.7091
(d) 376.3463

470. Tina is weighing objects in kilograms. A book weighs 0.923 kilograms; a pencil weighs 0.029 kilograms; an eraser weighs 0.1153 kilograms. What is the total weight of the three objects?
(a) 0.4353 kgs
(b) 1.0673 kgs
(c) 1.4283 kgs
(d) 10.673 kgs

471. The team played three basketball games last week. Monday's game lasted 113.9 minutes; Wednesday's game lasted 106.7 minutes; Friday's game lasted 122 minutes. What is the average time, in minutes, for the three games?
(a) 77.6
(b) 103.2
(c) 114.2
(d) 115.6

472. Indra has two pieces of balsa wood. Piece A is 0.724 centimetres thick. Piece B is 0.0076 centimetres thicker than Piece A. How thick is Piece B?
(a) 0.7164
(b) 0.7316
(c) 0.8
(d) 0.08

1000 MATH PROBLEMS >>> Decimals

473. Mahesh has a twenty-rupees bill and a five Rupee coin in his wallet and Rs. 1.29 in change in his pocket. If he buys a half-gallon of ice cream that costs Rs. 4.89, how much money will he has left?
(a) Rs. 22.48
(b) Rs. 22.30
(c) Rs. 21.48
(d) Rs. 21.40

474. The butcher at Meat Market wants to divide ground beef into 8 packages. If each package weighs 0.75 kgs and he has 0.04 kgs of ground beef left over, how many kgs of ground beef did he start with?
(a) 5.064
(b) 5.64
(c) 6.04
(d) 6.4

475. It is 19.85 miles from Jenny's home to her job. If she works 5 days a week and drives to work, how many miles does Jenny drive each week?
(a) 99.25
(b) 188.5
(c) 190.85
(d) 198.5

476. Puru and Ashi went out to dinner and spent a total of Rs. 42.09. If they tipped the waiter Rs. 6.25 and the tip was included in their total bill, how much did their meal alone cost?
(a) Rs. 35.84
(b) Rs. 36.84
(c) Rs. 36.74
(d) Rs. 48.34

477. Arun earns Rs. 8.30 an hour for the first 40 hours he works each week. For every hour he works overtime, he earns 1.5 times his regular hourly wage. If Arun worked 44 hours last week, how much money did he earn?
(a) Rs. 365.20
(b) Rs. 337.50
(c) Rs. 381.80
(d) Rs. 547.80

478. The highest temperature in Spring Valley on September 1 was 93.6°F. On September 2, the highest temperature was 0.8 degrees higher than on September 1. On September 3, the temperature was 11.6 degrees lower than on September 2. What was the temperature on September 3?
(a) 74°F
(b) 82.2°F
(c) 82.8°F
(d) 90°F

479. A survey has shown that a family of four can save about Rs. 40 a week if they purchase generic items rather than brand-name ones. How much can a particular family save over 6 months? (1 month = 4.3 weeks)
(a) 1,032
(b) 1,320
(c) 1,310
(d) 1,300

480. The Benton High School girl's relay team ran the mile in 6.32 minutes in April. By May, they were able to run the same race in 6.099 minutes. By how many minutes had their time improved?
(a) 0.221
(b) 0.339
(c) 0.467
(d) 0.67

SECTION

4

PERCENTAGES

The following 11 sets of problems deal with percentages, which, like decimals, are a special kind of fraction. Percentages have many everyday uses, from figuring the tip in a restaurant to understanding complicated interest and inflation rates. This section will give you practice in working with the relationship between percents, decimals, and fractions, and with changing one into another. You'll also do a few ratios and proportions, which are a lot like percentages.

1000 MATH PROBLEMS >>> Percentages

SET 31

481. 2% =
 (a) 2.0
 (b) 0.2
 (c) 0.02
 (d) 0.002

482. 4% =
 (a) 0.04
 (b) 0.4
 (c) 4.0
 (d) 0.004

483. 25% =
 (a) 0.025
 (b) 0.25
 (c) 2.5
 (d) 2.05

484. 2.06% =
 (a) 2.06
 (b) 0.206
 (c) 0.0206
 (d) 0.00206

485. 400% =
 (a) 0.04
 (b) 0.4
 (c) 4.0
 (d) 40.0

486. 0.02 =
 (a) 0.20%
 (b) 2.0%
 (c) 20.0%
 (d) 200%

487. $6\frac{1}{4}\%$ =
 (a) 0.625%
 (b) 6.25%
 (c) 62.5%
 (d) 625%

488. 0.005% =
 (a) 0.00005
 (b) 0.0005
 (c) 0.005
 (d) 0.05

490. $\frac{1}{4}\%$ =
 (a) 0.0025%
 (b) 0.025%
 (c) 0.25%
 (d) 25.0%

490. $\frac{1}{4}$ =
 (a) 25%
 (b) 0.25%
 (c) 0.025%
 (d) 0.0025%

491. 0.97 is equal to
 (a) 97%
 (b) 9.7%
 (c) 0.97%
 (d) 0.097%

492. 10% converted to a fraction =
 (a) $\frac{100}{10}$
 (b) $\frac{1}{10}$
 (c) $\frac{10}{1}$
 (d) $\frac{10}{10}$

1000 MATH PROBLEMS >>> Percentages

493. 350% converted to a mixed number =

(a) $35\frac{1}{2}$

(b) $3\frac{1}{2}$

(c) $0.3\frac{1}{2}$

(d) 0.350

494. 24% converted to a fraction =

(a) $\frac{1}{24}$

(b) $\frac{6}{24}$

(c) $\frac{1}{25}$

(d) $\frac{6}{25}$

495. 80% of 400 =

(a) 480

(b) 340

(c) 320

(d) 180

496. 60% of 390 =

(a) 234

(b) 190

(c) 180

(d) 134

SET 32

497. 42% of 997 =
(a) 990.24
(b) 499.44
(c) 450.24
(d) 418.74

498. 300% of 20 =
(a) 7
(b) 20
(c) 30
(d) 60

499. 20% of 96 =
(a) 19.2
(b) 1.92
(c) 0.92
(d) 0.092

500. 26% converted to a decimal =
(a) 0.0026
(b) 0.026
(c) 0.26
(d) 2.6

501. Which of the following is 14 percent of 232?
(a) 3.248
(b) 32.48
(c) 16.57
(d) 165.7

502. What percentage of 18,000 is 234?
(a) 1,300%
(b) 130%
(c) 13%
(d) 1.3%

503. What percentage of 600 is 750?
(a) 80%
(b) 85%
(c) 110%
(d) 125%

504. What is $7\frac{1}{5}$% of 465, rounded to the nearest tenth?
(a) 32.5
(b) 33
(c) 33.5
(d) 34

505. 62.5% is equal to
(a) $\frac{1}{16}$
(b) $\frac{5}{8}$
(c) $6\frac{1}{4}$
(d) $6\frac{2}{5}$

506. Convert $\frac{7}{40}$ to a percentage.
(a) 0.0175%
(b) 0.175%
(c) 1.75%
(d) 17.5%

507. You can quickly figure a 20% tip on a restaurant bill of Rs. 18 by
(a) multiplying 18 × 20, then rounding down
(b) multiplying 18 × 20
(c) multiplying 18 × 2, then moving the decimal over one space to the left
(d) multiplying 18 × $\frac{1}{2}$ and rounding up

1000 MATH PROBLEMS >>> Percentages

508. You can quickly figure a 15% tip on a restaurant bill of Rs. 12 by

(a) multiplying Rs. 12 × 10 by moving the decimal to the left one space to get Rs. 1.20, then adding half that (or Rs. 0.60)

(b) dividing Rs. 12 ÷ 2 to get 6, then adding 6 + 6 to get 12 again, then moving the decimal over one place to the right

(c) dividing 15 ÷ Rs. 12, then rounding up

(d) dividing Rs. 12 ÷ 15, then rounding down

509. What is 43.4% of 15?

(a) 1.43
(b) 4.91
(c) 6.00
(d) 6.51

510. What is 0.2% of 20?

(a) 4.0
(b) 0.4
(c) 0.04
(d) 0.0044

511. What is 44% of 5?

(a) 0.22
(b) 2.2
(c) 2.02
(d) 0.0022

512. 80% is equivalent to

(a) 0.8 and $\frac{8}{10}$

(b) 8.0 and $\frac{8}{10}$

(c) 0.8 and $\frac{10}{8}$

(d) 0.08 and $\frac{8}{8}$

SET 33

513. 35% of what number is equal to 14?
    (a) 4
    (b) 40
    (c) 49
    (d) 400

514. Which of the following phrases means "percent"?
    (a) "per part"
    (b) "per 100 parts"
    (c) "per fraction"
    (d) "per decimal"

515. Which of the following phrases expresses a ratio?
    (a) "as to"
    (b) "out of"
    (c) "reduces to"
    (d) "convert from"

516. Which of the following is equal to 0.13?
    (a) 13
    (b) $\frac{13}{100}$
    (c) $\frac{13}{13}$
    (d) $\frac{100}{13}$

517. Which of the following expresses the meaning of the word "fraction"?
    (a) bottom number divided by top number
    (b) top number divided by bottom number
    (c) 100 divided by itself
    (d) multiply by 100

518. What percent of 50 is 12?
    (a) 4%
    (b) 14%
    (c) 24%
    (d) 0.4%

519. 33 is 12% of what number?
    (a) 3,300
    (b) 330
    (c) 275
    (d) 99

520. Change $\frac{4}{25}$ to a percent
    (a) 4%
    (b) 16%
    (c) 40%
    (d) 100%

521. 0.8 =
    (a) 8%
    (b) 0.8%
    (c) 80%
    (d) 800%

522. $\frac{1}{4} \times 100\%$ =
    (a) 25%
    (b) 0.25%
    (c) 2%
    (d) 2.5%

523. Membership dues at Arun's Gym are Rs. 53 per month this year, but were Rs. 50 per month last year. What was the percentage increase in the gym's prices?
    (a) 5.5%
    (b) 6.0%
    (c) 6.5%
    (d) 7.0%

1,000 MATH PROBLEMS >>> Percentages

524. Leela spent 15% of her allowance on clothes and the rest on CDs. If she spent Rs. 55 on clothes, how much did she spend on CDs?
(a) Rs. 311.60
(b) Rs. 314.50
(c) Rs. 323.60
(d) Rs. 350.50

525. Yogita just got a raise of $3\frac{1}{4}$%. Her original salary was Rs. 30,600. How much does she make now?
(a) Rs. 30,594.50
(b) Rs. 31,594.50
(c) Rs. 32,094.50
(d) Rs. 32,940.50

Answer question 526 on the basis of the following passage.

Basic cable television service, which includes 16 channels, costs Rs. 15 a month. The initial labor fee to install the service is Rs. 25. A Rs. 65 deposit is required but will be refunded within two years if the customer's bills are paid in full. Other cable services may be added to the basic service: the movie channel service is Rs. 9.40 a month; the news channels are Rs. 7.50 a month; the arts channels are Rs. 5.00 a month; the sports channels are Rs. 4.80 a month.

526. A customer's cable television bill totaled Rs. 20 a month. Using the passage above, what portion of the bill was for basic cable service?
(a) 25%
(b) 33%
(c) 50%
(d) 75%

527. Geeta has finished reading 30% of a 340-page novel. How many pages has she read?
(a) 102
(b) 103
(c) 105
(d) 113

528. Ten students from the 250-student senior class at St. Martin High School have received full college scholarships. What percentage of the senior class received full college scholarships?
(a) 2%
(b) 4%
(c) 10%
(d) 25%

SET 34

529. Krishna's dinner at a local restaurant cost Rs. 13.85. If she wants to leave the server a tip that equals 20% of the cost of her dinner, how much of a tip should she leave?

(a) Rs. 2.00
(b) Rs. 2.67
(c) Rs. 2.77
(d) Rs. 3.65

530. This month, attendance at the baseball park increased 150% over what it was last month. If attendance this month was 280,000, what was the attendance last month, rounded to the nearest whole number?

(a) 140,000
(b) 176,670
(c) 186,667
(d) 205,556

531. Manish purchased a house for Rs. 70,000. Five years later, he sold it for an 18% profit. What was his selling price?

(a) Rs. 82,600
(b) Rs. 83,600
(c) Rs. 85,500
(d) Rs. 88,000

532. The price of gasoline drops from Rs. 1.00 per gallon to Rs. 0.95 per gallon. What is the percent of decrease?

(a) 2%
(b) 3%
(c) 4%
(d) 5%

533. A certain power company gives a $1\frac{1}{2}\%$ discount if a customer pays the bill at least ten days before the due date. If Teena pays for Rs. 48.50 bill ten days early, how much money will she save, rounded to the nearest paise?

(a) Rs. 0.49
(b) Rs. 0.73
(c) Rs. 1.50
(d) Rs. 7.28

534. Each year, on average, 4 of the 1600 prisoners who pass through the county jail escape. What percentage of prisoners escapes?

(a) 16%
(b) 4%
(c) 0.04%
(d) 0.25%

535. Mona and Philip arrive late for a movie and miss 10% of it. The movie is 90 minutes long. How many minutes did they miss?

(a) 10 minutes
(b) 9 minutes
(c) 8 minutes
(d) 7 minutes

536. Tony ate 3 ounces of a 16-ounce carton of ice cream. What percentage of the carton did he eat?

(a) 18.75%
(b) 17.25%
(c) 19.50%
(d) 16.75%

537. Kirti's pay is Rs. 423.00; 19 percent of that is subtracted for taxes. How much is her take-home pay?
(a) Rs. 404.44
(b) Rs. 355.46
(c) Rs. 342.63
(d) Rs. 455.45

538. In the electrical engineering department, $66\frac{2}{3}\%$ of the students are women and 100 of the students are men. How many electrical engineering students are there in all?
(a) 400
(b) 200
(c) 300
(d) 500

539. A ticket to an evening movie at the Bharat costs Rs. 7.50. The cost of popcorn at the concession stand is 80% of the cost of a ticket. How much does popcorn cost?
(a) Rs. 6.00
(b) Rs. 6.50
(c) Rs. 7.00
(d) Rs. 5.50

540. This week the stock market closed at 795. Last week it closed at 644. What was the percentage of increase this week, rounded to the closest whole number?
(a) 80%
(b) 81%
(c) 82%
(d) 83%

541. Veena borrowed Rs. 10,000.00 from her Uncle Kiran and agreed to pay him 4.5% interest, compounded yearly. How much interest did she owe the first year?
(a) Rs. 145.00
(b) Rs. 100.00
(c) Rs. 400.00
(d) Rs. 450.00

542. Textile officers have to buy duty shoes at the full price of Rs. 84.50, but officers who have served at least a year get a 15% discount. Officers who have served at least three years get an additional 10% off the discounted price. How much does an officer who has served at least three years have to pay for shoes?
(a) Rs. 63.78
(b) Rs. 64.65
(c) Rs. 71.83
(d) Rs. 72.05

543. At the city park, 32% of the trees are oaks. If there are 400 trees in the park, how many trees are NOT oaks?
(a) 128
(b) 272
(c) 278
(d) 312

544. The town of Delhi spends 15% of its annual budget on its public library. If Delhi spent Rs. 3,000 on its public library this year, what was its annual budget this year?
(a) Rs. 15,000
(b) Rs. 20,000
(c) Rs. 35,000
(d) Rs. 45,000

SET 35

545. Of the 1,200 videos available for rent at a certain video store, 420 are comedies. What percent of the videos are comedies?

(a) $28\frac{1}{2}\%$
(b) 30%
(c) 32%
(d) 35%

546. Tony, a Golden Retriever, gained 5.1 kgs this month. If Tony now weighs 65.1 kgs, what is the percent increase in Tony's weight?

(a) 5.9%
(b) 6%
(c) 8.5%
(d) 9.1%

547. Giri bought a sofa-sleeper at a 10% off sale and paid the sale price of Rs. 575.00. What was the price, rounded to the nearest cent, of the sofa-sleeper before the sale?

(a) Rs. 585.00
(b) Rs. 587.56
(c) Rs. 633.89
(d) Rs. 638.89

548. Naveen saves $5\frac{1}{4}\%$ of his weekly salary. If Naveen earns Rs. 380.00 per week, how much does he save each week?

(a) Rs. 19.95
(b) Rs. 20.52
(c) Rs. 21.95
(d) Rs. 25.20

549. Usha has completed 70% of her homework. If she has been doing homework for 42 minutes, how many more minutes does she have left to work?

(a) 15
(b) 18
(c) 20.5
(d) 28

550. Tarun cuts a piece of rope into three pieces. One piece is 8 feet long, one piece is 7 feet long, and one piece is 5 feet long. The shortest piece of rope is what percent of the original length before the rope was cut?

(a) 4%
(b) 18.5%
(c) 20%
(d) 25%

551. Romy's monthly food budget is equal to 40% of her monthly house payment. If her food budget is Rs. 200.00 a month, how much is her house payment each month?

(a) Rs. 340.00
(b) Rs. 400.00
(c) Rs. 500.00
(d) Rs. 540.00

552. If container A holds 8 gallons of water, and container B holds 12% more than container A, how many gallons of water does container B hold?

(a) 8.12
(b) 8.48
(c) 8.96
(d) 9

1000 MATH PROBLEMS >>> Percentages

553. Thirty-five paise is what percent of Rs. 1.40?

(a) 25
(b) 40
(c) 45
(d) 105

554. The Colgate Toothbrush Company hired 30 new employees. This hiring increased the company's total workforce by 5%. How many employees now work at Colgate?

(a) 530
(b) 600
(c) 605
(d) 630

555. Mohan had 200 baseball cards. He sold 5% of the cards on Saturday and 10% of the remaining cards on Sunday. How many cards are left?

(a) 170
(b) 171
(c) 175
(d) 185

556. Kiran spent 45% of the money that was in her wallet. If she spent Rs. 9.50, how much money was in her wallet to begin with? Round to the nearest paise.

(a) Rs. 4.28
(b) Rs. 4.38
(c) Rs. 14.49
(d) Rs. 21.11

557. Suraj earned a $4\frac{3}{4}\%$ pay raise. If his salary was Rs. 27,400 before the raise, how much was his salary after the raise?

(a) Rs. 27,530.15
(b) Rs. 28,601.50
(c) Rs. 28,701.50
(d) Rs. 29,610.50

558. The temperature in Sun village reached 100 degrees or more about 15% percent of the past year. About how many days did the temperature in Sun Village climb to 100 or climb to 100 or more? (1 year = 365 days) Round your answer.

(a) 45
(b) 54
(c) 55
(d) 67

559. A certain radio station plays classical music during 20% of its airtime. If the station is on the air 24 hours a day, how many hours each day is the station NOT playing classical music?

(a) 8
(b) 15.6
(c) 18.2
(d) 19.2

560. In order to pass a certain exam, candidates must answer 70% of the test questions correctly. If there are 70 questions in the exam, how many questions must be answered correctly in order to pass?

(a) 49
(b) 52
(c) 56
(d) 60

1000 MATH PROBLEMS >>> *Percentages*

SET 36

561. Of 150 people polled, 105 said they rode the city bus at least 3 times per week. How many people out of 100,000 could be expected to ride the city bus at least 3 times each week?
(a) 55,000
(b) 70,000
(c) 72,500
(d) 75,000

562. In a given area of the United States, in one year, there were about 215 highway accidents associated with drinking alcohol. Of these, 113 were caused by excessive speed. About what percent of the accidents were speed-related?
(a) 47%
(b) 49%
(c) 51%
(d) 53%

563. Reema earns Rs. 26,000 a year. If she receives a 4.5% salary increase, how much will she earn?
(a) Rs. 26,450
(b) Rs. 27,170
(c) Rs. 27,260
(d) Rs. 29,200

564. A sprinkler system installed in a home that is under construction will cost about 1.5% of the total building cost. The same system, installed after the home is built, is about 4% of the total building cost. How much would a homeowner save by installing a sprinkler system in Rs. 150,000 home while the home is still under construction?
(a) Rs. 600
(b) Rs. 2,250
(c) Rs. 3,750
(d) Rs. 6,000

565. Out of 100 shoppers polled, 80 said they buy fresh fruit every week. How many shoppers out of 30,000 could be expected to buy fresh fruit every week?
(a) 2,400
(b) 6,000
(c) 22,000
(d) 24,000

566. Nationwide, in one year there were about 21,500 residential fires associated with furniture. Of these, 11,350 were caused by smoking materials. About what percent of the residential fires were smoking-related?
(a) 47%
(b) 49%
(c) 50%
(d) 53%

567. A locked ammunition box is about $2\frac{1}{2}$ centimetres thick. About how thick is this box in inches? (1 cm = 0.39 inches)
(a) $\frac{1}{4}$ inch
(b) 1 inch
(c) 2 inches
(d) 5 inches

568. A pump installed on a well can pump at a maximum rate of 100 gallons per minute. If the town requires 15,000 gallons of water in one day, how long would the pump have to continuously run at 75% of its maximum rate to meet the town's need?
(a) 112.5 minutes
(b) 150 minutes
(c) 200 minutes
(d) 300 minutes

1000 MATH PROBLEMS >>> Percentages

569. A company makes several items, including filing cabinets. One-third of their business consists of filing cabinets, and 60% of their filing cabinets are sold to businesses. What percent of their total business consists of filing cabinets sold to businesses?

(a) 20%
(b) 33%
(c) 40%
(d) 60%

570. A gram of fat contains 9 calories. An 1,800-calorie diet allows no more than 20% calories from fat. How many grams of fat are allowed in that diet?

(a) 40 g
(b) 90 g
(c) 200 g
(d) 360 g

571. Mr. Jain's temperature is 98 degrees Fahrenheit. What is his temperature in degrees Celsius?

$C = \frac{5}{9}(F - 32)$

(a) 35.8
(b) 36.7
(c) 37.6
(d) 31.1

572. If no treatment has been given within 3 hours after injury to a certain organ, the organ's function starts decreasing by 20% each hour. If no treatment has been given within 6 hours after injury, approximately how much function will remain?

(a) 51.2%
(b) 60%
(c) 70%
(d) 80%

573. Of 1,125 nurses who work in the hospital, 135 speak fluent Hindi. What percentage of the nursing staff speaks fluent Hindi?

(a) 7.3%
(b) 8.3%
(c) 12%
(d) 14%

574. The basal metabolic rate (BMR) is the rate at which our body uses calories. The BMR for a man in his twenties is about 1,700 calories per day. If 204 of those calories should come from protein, about what percent of this man's diet should be protein?

(a) 1.2%
(b) 8.3%
(c) 12%
(d) 16%

575. The condition Down syndrome occurs in about 1 in 1,500 children when the mothers are in their twenties. About what percent of all children born to mothers in their twenties are likely to have Down syndrome?

(a) 0.0067%
(b) 0.067%
(c) 0.67%
(d) 6.7%

576. A 15 cc dosage must be increased by 20%. What is the new dosage?

(a) 17 cc
(b) 18 cc
(c) 30 cc
(d) 35 cc

SET 37

577. During exercise, a person's heart rate should be between sixty and ninety percent of the difference between 220 and the person's age. According to this guideline, what should a 30-year-old person's maximum heart rate be during exercise?
   (a) 114
   (b) 132
   (c) 171
   (d) 198

578. Of 9,125 patients treated in a certain emergency room in one year, 72% were male. Among the males, 3 out of 5 were under age 25. How many of the emergency room patients were males age 25 or older?
   (a) 2,628
   (b) 3,942
   (c) 5,475
   (d) 6,570

579. A certain water pollutant is unsafe at a level of 20 ppm (parts per million). A city's water supply now contains 50 ppm of this pollutant. What percentage improvement will make the water safe?
   (a) 30%
   (b) 40%
   (c) 50%
   (d) 60%

580. In half of migraine sufferers, a certain drug reduces the number of migraines by 50%. What percentage of all migraines can be eliminated by this drug?
   (a) 25%
   (b) 50%
   (c) 75%
   (d) 100%

581. There are 176 men and 24 women serving the 9th Precinct. What percentage of the 9th Precinct's force is women?
   (a) 12%
   (b) 14%
   (c) 16%
   (d) 24%

Use the pie chart to answer question 582

582. The chart shows quarterly sales for Cool-Air's air-conditioning units. Which of the following combinations contributed 70% to the total?

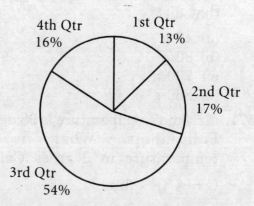

Sales For 1997

   (a) first and second quarters
   (b) second and third quarters
   (c) second and fourth quarters
   (d) third and fourth quarters

583. In the James school district last year, 220 students were vaccinated for measles, mumps, and rubella. Of those, 60 percent reported that they had flu at some time in their lives. How many students had not had the flu previously?
   (a) 36
   (b) 55
   (c) 88
   (d) 126

1000 MATH PROBLEMS >>> Percentages

584. A floor plan is drawn to scale so that one quarter inch represents 2 feet. If a hall on the plan is 4 inches long, how long will the actual hall be when it is built?

(a) 2 feet
(b) 8 feet
(c) 16 feet
(d) 32 feet

585. A doctor can treat 4 heart patients per hour; however, stroke patients need 3 times as much of the doctor's time. If the doctor treats patients 6 hours per day and has already treated 10 heart patients and 3 stroke patients today, how many more stroke patients will she have time to treat today?

(a) 1
(b) 2
(c) 3
(d) 4

586. An insurance policy pays 80% of the first Rs. 20,000 of a certain patient's medical expenses, 60% of the next Rs. 40,000 and 40% of the Rs. 40,000 after that. If the patient's total medical bill is Rs. 92,000, how much will the policy pay?

(a) Rs. 36,800
(b) Rs. 49,600
(c) Rs. 52,800
(d) Rs. 73,600

587. A pump installed on a well can pump at a maximum rate of 100 gallons per minute. If the pump runs at 50% of its maximum rate for six hours a day, how much water is pumped in one day?

(a) 3.00×10^2 gallons
(b) 1.80×10^4 gallons
(c) 3.60×10^4 gallons
(d) 7.20×10^4 gallons

588. Meera can grade five of her student's papers in an hour. Joe can grade four of the same papers in an hour. If Meera works for three hours grading, and Joe works for two hours, what percentage of the 50 student's papers will be graded?

(a) 44%
(b) 46%
(c) 52%
(d) 54%

589. Dinesh has 40 math problems to do for homework. If he does 40% of the assignment in one hour, how long will it take for Dinesh to complete the whole assignment?

(a) 1.5 hours
(b) 2.0 hours
(c) 2.5 hours
(d) 3.0 hours

1000 MATH PROBLEMS >>> Percentages

590. The population of Smithtown increases at a rate of 3% annually. If the population is currently 2,500, what will the population be at the same time next year?

(a) 2,530
(b) 2,560
(c) 2,575
(d) 2,800

591. The Kapoor family travelled 75 miles to visit relatives. If they travelled 57.8% of the way before they stopped at a gas station, how far was the gas station from their relatives' house? Round your answer to the nearest $\frac{1}{3}$ mile.

(a) $31\frac{2}{3}$ miles
(b) $32\frac{2}{3}$ miles
(c) 35 miles
(d) $38\frac{1}{3}$ miles

592. A recent survey polled 2,500 people about their reading habits. The results are as follows:

READING SURVEY

Books per month	Percentage
0	13
1-3	27
4-6	32
>6	28

How many people surveyed had read books in the last month?

(a) 700
(b) 1,800
(c) 1,825
(d) 2,175

1000 MATH PROBLEMS >>> Percentages

SET 38

593. A machine on a production line produces parts that are not acceptable by company standards four percent of the time. If the machine produces 500 parts, how many will be defective?

(a) 8
(b) 10
(c) 16
(d) 20

594. Thirty percent of the high school is involved in athletics. If 15% of the athletes play football, what percentage of the whole school plays football?

(a) 4.5%
(b) 9.0%
(c) 15%
(d) 30%

595. Twenty percent of the people at a restaurant selected the dinner special. If 40 people did not select the special, how many people are eating at the restaurant?

(a) 10
(b) 20
(c) 40
(d) 50

596. An average of 90% is needed on five tests to receive an A in a class. If a student received scores of 95, 85, 88 and 84 on the first four tests, what will the student need to get an A?

(a) 92
(b) 94
(c) 96
(d) 98

Answer questions 597 and 598 on the basis of the diagram below.

CURRENT FIRE STATISTICS

Cause	Fires (% of Total)	Civilian Deaths (% of Total)
Heating equipment	161,500 (27.5%)	770 (16.8%)
Cooking equipment	104,800 (17.8%)	350 (7.7%)
Incendiary, suspicious	65,400 (11.1%)	620 (13.6%)
Electrical equipment	45,700 (7.8%)	440 (9.6%)
Other equipment	43,000 (7.3%)	240 (5.3%)
Smoking materials	39,300 (6.7%)	1,320 (28.9%)
Appliances, air conditioning	36,200 (6.2%)	120 (2.7%)
Exposure and other heat	28,600 (4.8%)	191 (4.2%)
Open flame	27,200 (4.6%)	130 (2.9%)
Child play	26,900 (4.6%)	370 (8.1%)
Natural causes	9,200 (1.6%)	10 (0.2%)

597. What is the percentage of the total fires caused by electrical equipment and other equipment combined?

(a) 7.8%
(b) 14.9%
(c) 15.1%
(d) 29.9%

598. Of the following causes, which one has the highest ratio of total fires to percentage of deaths?

(a) heating equipment
(b) smoking materials
(c) exposure and other heat
(d) child play

1000 MATH PROBLEMS >>> Percentages

599. The City Bus Department operates 200 bus routes. Of these, $5\frac{1}{2}\%$ are express routes. How many express routes are there?
(a) 11
(b) 15
(c) 22
(d) 25

600. A study shows that 600,000 women die each year in pregnancy and childbirth, one-fifth more than scientists previously estimated. How many such deaths did the scientists previously estimate?
(a) 120,000
(b) 300,000
(c) 480,000
(d) 500,000

601. A certain baseball player gets a hit about 3 out of every 12 times he is at bat. What percentage of the time he is at bat does he get a hit?
(a) 25%
(b) 32%
(c) 35%
(d) 40%

602. Harish has worked 40% of his 8-hour shift at the widget factory. How many hours has he worked?
(a) 3 hours
(b) 3.2 hours
(c) 3.4 hours
(d) 3.5 hours

603. 15 of the 34 guests at the Animal Refuge League banquet ordered vegetarian dinners. What percent of the guests ordered vegetarian meals?
(a) 22%
(b) 40%
(c) 44%
(d) 49%

604. Satish estimates that his college expenses this year will be Rs. 26,000, however, he wishes to add 15% to that amount in case of emergency. How much should Satish try to add to his college fund?
(a) Rs. 1,733
(b) Rs. 2,900
(c) Rs. 3,000
(d) Rs. 3,900

605. Because it was her birthday, Patty spent 325% more than she usually spends for lunch. If she usually spends Rs. 4.75 per day for lunch, how much did she spend today?
(a) Rs. 9.50
(b) Rs. 15.44
(c) Rs. 20.19
(d) Rs. 32.50

606. Hemant weighed 430 kilos in 1996. Today he weighs 337 kilo. About what percentage of weight has he lost since 1996?
(a) 78%
(b) 30%
(c) 22%
(d) 15%

1000 MATH PROBLEMS >>> Percentages

607. Ramesh, a TV repairperson, gives a $5\frac{1}{2}\%$ discount to senior citizens. If Hanif, age 73, owes a repair bill of Rs. 75, what is the amount of the final bill after Ramesh applies the discount?

(a) Rs. 34.88
(b) Rs. 50.48
(c) Rs. 70.88
(d) Rs. 73.88

608. Pradip took a 250-question physics test and got 94% of the answers correct. How many questions did he answer correctly?

(a) 240 questions
(b) 235 questions
(c) 230 questions
(d) 225 questions

SET 39

609. A painting by a famous artist, Mr. Sam, is sold for Rs. 875. Sam's agent gets 15% of the proceeds and the gallery where the painting was sold gets 27%. How much does Sam have left?

(a) Rs. 236.25
(b) Rs. 367.5
(c) Rs. 507.5
(d) Rs. 607.5

610. After paying his agent a commission of 15%, Congressperson Rajnish is left with Rs. 332,000 as advance payment for his book on political campaigning techniques. What was the amount of his original advance?

(a) Rs. 49,800
(b) Rs. 381,800
(c) Rs. 390,588
(d) Rs. 282,200

611. At the Mona's Boutique 8% of the dresses are designer dresses, the rest are fakes. If there are 300 dresses at the boutique, how many are NOT designer dresses?

(a) 136
(b) 276
(c) 296
(d) 292

612. Amit spends 25% of his weekly budget on chocolate snack cakes. This week he spent Rs. 12 on snack cakes. What was his weekly budget this week?

(a) Rs. 3
(b) Rs. 48
(c) Rs. 36
(d) Rs. 18

613. There are 26 pies in the country fair pie contest this year. Of these, 4 are peach. What percent of the pies are peach?

(a) 4%
(b) 9%
(c) 12%
(d) 15%

614. Mona, a 7-year old entrepreneur, earned Rs. 5 this week selling peanut butter sandwiches to kids whose mothers had packed them yucky lunches. If Mona now has a total of Rs. 40 from her sale of sandwiches, what was the percent increase in Mona's sandwich fund this week over last week's total?

(a) 4.5%
(b) 8.5%
(c) 14.3%
(d) 15.5%

615. A 15% mark-up on merchandise originally priced at Rs. 26.50 would be

(a) Rs. 2.97
(b) Rs. 3.50
(c) Rs. 3.67
(d) Rs. 3.97

616. Manish bought a used chain saw at 20% off and paid Rs. 375.00 What was the price of the chain saw before it was marked down?

(a) Rs. 485.00
(b) Rs. 465.56
(c) Rs. 468.75
(d) Rs. 478.89

1000 MATH PROBLEMS >>> Percentages

617. Babita buys six dolls and saves $3\frac{1}{2}\%$ of the total price by buying in bulk. If each doll originally costs Rs. 300, how much does Barbara save?

(a) Rs. 10.50
(b) Rs. 54.00
(c) Rs. 63.00
(d) Rs. 75.00

618. Manish, a motel housekeeper, has finished cleaning about 40% of the 32 rooms he's been assigned. About how many more rooms does he have left to clean?

(a) 29
(b) 25
(c) 21
(d) 19

619. Girish cuts a piece of string cheese into three pieces. One piece is 6 inches long, one piece is 4 inches long, and one piece is 3 inches long. The shortest piece of string cheese is approximately what percent of the original length before the string cheese was cut?

(a) 30%
(b) 27%
(c) 23%
(d) 20%

620. Ravi's monthly gambling debt payment is equal to 70% of his monthly food budget, which is Rs. 450. How much is Ravi's gambling debt payment each month?

(a) Rs. 70
(b) Rs. 315
(c) Rs. 325
(d) Rs. 400

621. Forty paise is what percent of Rs. 1.30?

(a) 40%
(b) 31%
(c) 20%
(d) 11%

622. In the City Zoo the ratio of male to female Komodo dragons is 3 to 5. What percentage of male Komodo dragons won't have a date on Saturday night?

(a) 60%
(b) 32%
(c) 15%
(d) 6%

623. Last year in our town, the ratio of rainy days to sunny ones was 2 to 5. How many rainy days did we have? (1 year = 365 days)

(a) 56
(b) 76
(c) 106
(d) 146

624. Sam had a bag of 150 cookies. He ate 4% of the cookies while watching cartoons on Saturday morning and 15% of the remaining cookies while watching detective show re-runs on Saturday afternoon. About how many cookies did he have left?

(a) 144
(b) 122
(c) 28
(d) 22

SET 40

625. Hena and Jenny leave from different points walking towards each other. Hena walks $2\frac{1}{2}$ miles per hour and Jenny walks 4 miles per hour. If they meet in $2\frac{1}{2}$ hours, how far apart were they?

(a) 9 miles
(b) 13 miles
(c) $16\frac{1}{4}$ miles
(d) $18\frac{1}{2}$ miles

626. 30% apples are rotten. How many apples are there in the basket if in a basket 135 apples are rotten?

(a) 515 apples
(b) 450 apples
(c) 405 apples
(d) 425 apples

627. Lucy and Leela dropped one of every eight chocolates from the conveyer belt onto the floor. What percentage of chocolates did they drop on the floor?

(a) 12.5%
(b) 13.2%
(c) 14.5%
(d) 15.2%

628. Pradip gave the pizza delivery person a tip of Rs. 4.00, which was 20% of his total bill for the pizza he ordered. How much did the pizza cost?

(a) Rs. 30.00
(b) Rs. 25.00
(c) Rs. 15.00
(d) Rs. 20.00

629. A popular news show broadcasts for an hour. During that time, there are 15 minutes of commercials. What percentage of the hour is devoted to commercials?

(a) 15%
(b) 35%
(c) 25%
(d) 45%

630. Minoo had 75 stuffed animals. Her grandmother gave fifteen of them to her. What percentage of the stuffed animals did her grandmother give her?

(a) 20%
(b) 15%
(c) 25%
(d) 10%

631. At the garage, Meena's bill to have her car repaired was Rs. 320.00. The total charge for labour was Rs. 80.00. What percentage of the bill was for labour?

(a) 15%
(b) 20%
(c) 25%
(d) 30%

632. Ram has completed 78% of his 200-page thesis. How many pages has he written?

(a) 150 pages
(b) 156 pages
(c) 165 pages
(d) 160 pages

1000 MATH PROBLEMS >>> Percentages

633. A house is valued at Rs. 185,000 in a community that assesses property at 85% of value. If the tax rate is Rs. 24.85 per thousand rupees assessed, how much is the property tax bill?

(a) Rs. 1,480
(b) Rs. 1,850
(c) Rs. 3,907.66
(d) Rs. 4,597.25

634. Ramesh secured 50% marks in English, 60% in Mathematics, 75% in Hindi. If the maximum marks in these subjects were 50, 70 and 80 respectively, find his aggregate percentage.

(a) 75.5%
(b) 65.25%
(c) 63.5%
(d) 57.5%

635. After paying a commission to his broker of 7% of the sale price, a seller receives Rs. 103,000 for his house. How much did the house sell for?

(a) Rs. 95,790
(b) Rs. 110,000
(c) Rs. 110,420
(d) Rs. 110,753

636. Sunil, a salesperson associated with broker Bobby, lists a house for Rs. 115,000, with 6% commission due at closing, and finds a buyer. Bobby's practice is that 45% of commissions go to his office and the rest to the salesperson. How much will Sunil make on the sale?

(a) Rs. 3,105
(b) Rs. 3,240
(c) Rs. 3,795
(d) Rs. 3,960

637. A storekeeper leases her store building for the following amount: Rs. 1,000 per month rent, $\frac{1}{12}$ of the Rs. 18,000 annual tax bill, and 3% of the gross receipts from her store. If the storekeeper takes in Rs. 75,000 in one month, what will her lease payment be?

(a) Rs. 4,750
(b) Rs. 3,750
(c) Rs. 3,250
(d) Rs. 7,400

638. The appraised value of a property is Rs. 325,000, assessed at 90% of its appraisal. If the tax rate for the year is Rs. 2.90 per thousand of assessment, how much are the taxes for the first half of the year?

(a) Rs. 471.25
(b) Rs. 424.13
(c) Rs. 942.50
(d) Rs. 848.25

1000 MATH PROBLEMS >>> Percentages

639. In 1978-79 India produced 132 million tonnes of foodgrains and in 1983-84 the production was 151.8 million tonnes. Find the increase percentage of the production of food grains.
(a) 15%
(b) 18%
(c) 20%
(d) 25%

640. In October-2000 the Govt. of India increased the price of petrol by 25%. By how much percent Satish should reduce his consumption of petrol. So that his expenditure on petrol does not increase?
(a) 20%
(b) 15%
(c) 25%
(d) 10%

SECTION

5

ALGEBRA

Basic algebra problems, such as the following 10 sets, ask you to solve equations in which one element, or more than one, is unknown and generally indicated by a letter of the alphabet (often either x or y). In doing the following problems, you will get practice in isolating numbers on one side of the equation and unknowns on the other, thus finding the replacement for the unknown. You'll also practice expressing a problem in algebraic form, particularly when you get to the word problems. Other skills you'll practice include working with exponents and roots, factoring, and dealing with polynomial expressions.

SET 41

641. Which of the following is the synonym for "equation"?
(a) additional units
(b) unknown quantity
(c) solve for
(d) balance scale

642. What percentage of 50 is 12?
(a) 4%
(b) 14%
(c) 24%
(d) 34%

643. If $8n + 25 = 65$, then n is
(a) 5
(b) 10
(c) 40
(d) 90

644. The sum of a number and its double is 69. What is the number?
(a) 46.6
(b) 34.5
(c) 23
(d) 20

645. Twelve less than 4 times a number is 20. What is the number?
(a) 2
(b) 4
(c) 6
(d) 8

646. A certain number when added to 50% of itself is 27. What is the number?
(a) 7
(b) 9
(c) 11
(d) 18

647. One-sixth of a certain number is four more than one-twelfth the number. Find the number.
(a) 6
(b) 18
(c) 36
(d) 48

648. Six less than $\frac{1}{9}$ of 45 is
(a) –1
(b) –2
(c) 1
(d) 3

649. Twelve times one-half of a number is thirty-six. What is the number?
(a) 3
(b) 6
(c) 8
(d) 18

650. 21 times four times one-twelfth of a number is seven. What is the number?
(a) 12
(b) 7
(c) 0.69
(d) 1

651. When both seven and three are taken away from a number the result is 31. What is the number?
(a) 20
(b) 36
(c) 41
(d) 55

1000 MATH PROBLEMS >>> *Algebra*

652. 35 is what percent of 90?
  (a) 0.25%
  (b) 0.38%
  (c) 25%
  (d) 38%

653. Six less than three times a number is four more than twice the number. Find the number.
  (a) 44
  (b) 2
  (c) 1
  (d) 10

654. The product of 16 and one-half a number is 136. Find the number.
  (a) 84
  (b) 16
  (c) 76
  (d) 17

655. 12 more than 30 percent of a number is one-half the number. Find the number.
  (a) 4
  (b) 18
  (c) 60
  (d) 72

656. Two times a number is the result when 7 times a number is taken away from 99. Find the number.
  (a) 85
  (b) 11
  (c) 46
  (d) 9

SET 42

657. Forty-five together with nine-fifths of a number is twice the number. What is the number?
(a) 23.4
(b) 162
(c) 26
(d) 225

658. 33 is 12 percent of which of the following numbers?
(a) 3,960
(b) 396
(c) 275
(d) 2,750

659. Sixty percent of 770 ÷ 6 is equal to 7 times what number?
(a) 1
(b) 11
(c) 71
(d) 77

660. 19 more than a certain number is 63. What is the number?
(a) 14
(b) 44
(c) 58
(d) 82

661. A number is three times larger when 10 is added to it. What is the number?
(a) 33
(b) 7
(c) 13
(d) 5

662. Eleven and forty-one together are divided by a number. If the result is 13, what is the number?
(a) 2
(b) 4
(c) 10.5
(d) 20.5

663. Four times a number plus twenty is equal to 72. What is the number?
(a) 13
(b) 16
(c) 17
(d) 23

664. Fifty plus three times a number is 74. What is the number?
(a) 2
(b) 4
(c) 6
(d) 8

665. The product of two and four more than three times a number is 20. What is the number?
(a) 2
(b) 16
(c) 44
(d) 87

666. To solve for an unknown in an equation, you must always
(a) add it in
(b) subtract it from
(c) isolate it on one side
(d) eliminate the inequality

1000 MATH PROBLEMS >>> Algebra

667. What is the value of y when $x = 3$ and $y = 5 + 4x$?
(a) 6
(b) 9
(c) 12
(d) 17

668. The product of 16 and one-half a number is 136. Find the number.
(a) 84
(b) 16
(c) 76
(d) 17

669. 12 more than 30 percent of a number is one-half the number. Find the number.
(a) 4
(b) 18
(c) 60
(d) 72

670. Two times a number is the result when 7 times a number is taken away from 99. Find the number.
(a) 85
(b) 11
(c) 46
(d) 9

671. Which value of x will make this number sentence true? $x + 25 \leq 13$.
(a) −13
(b) −11
(c) 12
(d) 38

672. $\frac{x}{4} + \frac{3x}{4} =$
(a) $\frac{1}{2}x$
(b) $\frac{x^3}{4}$
(c) 1
(d) x

1000 MATH PROBLEMS >>> Alegbra

SET 43

673. Which of the following lists three consecutive even integers whose sum is 30?
    (a) 9, 10, 11
    (b) 8, 10, 12
    (c) 9, 11, 13
    (d) 10, 12, 14

674. If $x = 6$, $y = -2$, and $z = 3$, what is the value of the following expression? $\frac{xz - xy}{z^2}$
    (a) $-\frac{2}{3}$
    (b) $\frac{2}{3}$
    (c) $3\frac{1}{3}$
    (d) 5

675. If $\frac{2x}{16} = \frac{12}{48}$, what is x?
    (a) 2
    (b) 3
    (c) 4
    (d) 5

676. Which value of x will make the following inequality true?
    $3x - 14 \leq = 3$.
    (a) 4
    (b) 6
    (c) 8
    (d) 10

677. Which of the following is a simplification of
    $(x^2 + 4x + 4) \div (x + 2)$?
    (a) $x - 2x + 4$
    (b) $x + 4$
    (c) $x^2 + 3x + 2$
    (d) $x + 2$

678. Which of the following is equivalent to $x^2 + 3x$?
    (a) $x(x + 3)$
    (b) $2(x + 3)$
    (c) $(x + 3)^2$
    (d) $(x + 1)(x + 3)$

679. When twenty-three is added to a number ninety-nine is the result. What is the number?
    (a) 67
    (b) 76
    (c) 108
    (d) 122

680. Which of the following is equivalent to $2x(3xy + y)$?
    (a) $6(x^2)y + 2xy$
    (b) $6xy + 2xy$
    (c) $5x2y + 2x + y$
    (d) $3xy + 2x + y$

681. $x^2 - 4x + 4 \div x - 2 =$
    (a) $x + 2$
    (b) $x - 2$
    (c) $x^2 - 2x + 2$
    (d) $x^2 - 3x + 2$

1000 MATH PROBLEMS >>> Algebra

682. Which of the following is equivalent to $2y^2$?
  (a) $2(y + y)$
  (b) $2y(y)$
  (c) $y^2 + 2$
  (d) $y + y + y + y$

683. $x(3x^2 + y) =$
  (a) $4x^2 + xy$
  (b) $4x^2 + x + y$
  (c) $3x^3 + 2xy$
  (d) $3x^3 + xy$

684. $[(2x^3y)^3](4x^2y^2)$ is equivalent to which of the following?
  (a) $32x^{11}y^5$
  (b) $8x^{11}y^5$
  (c) $32x^{18}y^6$
  (d) $8x^{18}i^6$

685. An equation of the form $\frac{a}{b} = \frac{c}{d}$ is
  (a) an inequality
  (b) a variable
  (c) a proportion
  (d) a monomial

686. Which mathematical expression best describes the quotient of two numbers added to a third number?
  (a) $(x)(y) \div z$
  (b) $(x \div y)(z)$
  (c) $x \div y + z$
  (d) $(x + y) \div z$

687. If $\frac{x}{2} + \frac{x}{6} = 4$, what is x?
  (a) $\frac{1}{24}$
  (b) $\frac{1}{6}$
  (c) 3
  (d) 6

688. If $\frac{x}{54} = \frac{2}{9}$, then x is
  (a) 6
  (b) 12
  (c) 18
  (d) 108

SET 44

689. Solve for x in the following equation:

$\frac{1}{3}x + 3 = 8$

(a) 33
(b) 15
(c) 11
(d) 3

690. Solve the following equation for x:
$2x - 7 = 4$

(a) $-\frac{3}{2}$
(b) $\frac{3}{2}$
(c) $\frac{11}{2}$
(d) 22

691. What is the value of y when $x = 8$ and $y = (x^2 \div 4) - 2$?

(a) 8
(b) 14
(c) 16
(d) 18

692. If $\frac{1}{16} = \frac{x}{54}$, what is x?

(a) 3.375
(b) 3.5
(c) 4
(d) 4.5

693. Solve for x in the following equation:

$1.5x - 7 = 12.5$

(a) 29.25
(b) 19.5
(c) 13
(d) 5.5

694. Which of the following is the largest possible solution to the following inequality:

$\frac{1}{3}x - 3 \leq 5$

(a) $\frac{2}{3}$
(b) $\frac{8}{3}$
(c) 6
(d) 24

695. Which of these equations is shown on the graph below?

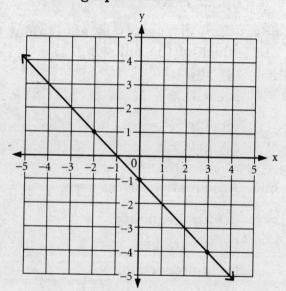

(a) $y = x - 1$
(b) $y = -1 - x$
(c) $y = 2x - 2$
(d) $y = 1 + x$

696. Evaluate the following expression if $a = 3$, $b = 4$, and $c = -2$:
$(ab - ac) \div abc$

(a) $-\frac{7}{8}$
(b) $-\frac{3}{4}$
(c) $-\frac{1}{4}$
(d) $\frac{1}{4}$

1000 MATH PROBLEMS >>> Algebra

697. If $\frac{x}{3} + \frac{x}{4} = 3$, what is x?

(a) $\frac{1}{12}$

(b) $\frac{7}{36}$

(c) $\frac{1}{4}$

(d) $5\frac{1}{7}$

698. A line passes through the points $(0, -1)$ and $(2, 3)$. What is the equation for the line?

(a) $y = \frac{1}{2}x - 1$

(b) $y = \frac{1}{2}x + 1$

(c) $y = 2x - 1$

(d) $y = 2x + 1$

699. When the product of three and a number is taken away from the sum of that number and six, the result is zero. What is the number?

(a) 3

(b) 7

(c) 9

(d) 14

700. Solve for x in the following equation:

$\frac{1}{3}x + 3 = 8$

(a) 33

(b) 15

(c) 11

(d) 3

701. The tens digit is four times the ones digit in a certain number. If the sum of the digits is 10, what is the number?

(a) 93

(b) 82

(c) 41

(d) 28

702. Which expression best describes the sum of three numbers multiplied by the sum of their reciprocals?

(a) $(a + b + c)(\frac{1}{a} + \frac{1}{b} + \frac{1}{c})$

(b) $(a)(\frac{1}{a}) + (b)(\frac{1}{b}) + (c)(\frac{1}{c})$

(c) $(a + b + c) \div (\frac{1}{a})(\frac{1}{b})(\frac{1}{c})$

(d) $(a)(b)(c) + (\frac{1}{a})(\frac{1}{b})(\frac{1}{c})$

703. Find the sum of $6x + 5y$ and $3x - 5y$.

(a) $9x + 10y$

(b) $9x - 10y$

(c) $10y$

(d) $9x$

704. 88 is the result when one-half of the sum of 24 and a number is all taken away from three times a number. What is the number?

(a) 16

(b) 40

(c) 56

(d) 112

SET 45

705. In algebra, a "variable" is
(a) the known quantity in the equation
(b) a symbol that stands for a number
(c) an inequality
(d) the solution for the equation

706. One third of the Length of a pole is under the ground. One-fourth in water and the remaining .5 metres above water. Find the total length of the pole.
(a) 8 m
(b) 14 m
(c) 10 m
(d) 12 m

707. A travelling circus can sell 250 admission tickets for Rs. 8 each. But if the tickets cost Rs. 6 each the circus can sell 400 tickets. How much larger are ticket sales when they cost Rs. 6 each than when they cost Rs. 8 each?
(a) Rs. 160
(b) Rs. 400
(c) Rs. 500
(d) Rs. 1700

708. Rs. 48 in tips is to be divided among three restaurant waiters. Tina gets three times more than Roma and Bina receives 4 times as much as Roma. How much does Bina receive?
(a) Rs. 18
(b) Rs. 16
(c) Rs. 24
(d) Rs. 6

709. The conversion from degrees Celsius to degrees Fahrenheit is $F = \frac{9}{5}C + 32$. If the temperature is 95 degrees Fahrenheit, what is the temperature in degrees Celsius to the nearest degree?
(a) 71 degrees
(b) 40 degrees
(c) 35 degrees
(d) 21 degrees

710. A family of three ate dinner at a restaurant, with a total bill of Rs. 240/-. If the mother's meal cost $\frac{5}{4}$ as much as the father's, and the child's meal was $\frac{3}{4}$ that of the father's how much was the father's meal?
(a) Rs. 60
(b) Rs. 70
(c) Rs. 80
(d) Rs. 100

711. Six kgs of a dried fruit mixture which costs 300 per kg and $1\frac{1}{2}$ kgs of nuts costing Rs 700 per kg are mixed together. What is the cost per kg of this mixture?
(a) Rs. 133
(b) Rs. 380
(c) Rs. 525
(d) Rs. 875

712. A helicopter flies over a river at 6:02 a.m. and arrives at a heliport 20 miles away at 6:17 a.m. How many miles per hour was the helicopter travelling?
(a) 120 mph
(b) 200 mph
(c) 30 mph
(d) 80 mph

1000 MATH PROBLEMS >>> Algebra

713. Two saline solutions are mixed. Twelve litres of 5% solution are mixed with 4 litres of 4% solution. What percent saline is the final solution?
    (a) 4.25%
    (b) 4.5%
    (c) 4.75%
    (d) 5%

714. Dinesh rides the first half of a bike race in two hours. If his partner Arun rides the return trip 5 miles per hour less, and it takes him three hours, how fast was Dinesh travelling?
    (a) 10 mph
    (b) 15 mph
    (c) 20 mph
    (d) 25 mph

715. Kiran can catch 10 fishes in an hour and Charan can catch five fishes in two hours. How long will Charan have to fish in order to catch the same number of fishes that Kiran would catch in two hours?
    (a) 2 hours
    (b) 4 hours
    (c) 6 hours
    (d) 8 hours

716. A grain elevator operator wants to mix two batches of corn with a resultant mix of 54 kgs per bushel. If he uses 20 bushels of 56 kgs per bushel corn, which of the following expressions gives the amount of 50 kgs per bushel corn needed?
    (a) $56x + 50x = 2x \times 54$
    (b) $20 \times 56 + 50x = (x + 20) \times 54$
    (c) $20 \times 56 + 50x = 2x \times 54$
    (d) $56x + 50x = (x + 20) \times 54$

717. Hira and Lina are both salespeople at a certain electronics store. If they made 36 sales one day, and Lina sold three less than twice Hira's sales total, how many units did Hira sell?
    (a) 19
    (b) 15
    (c) 12
    (d) 13

718. It will take Giri 4 days to string a certain fence. If Mini could string the same fence in 3 days, how long will it take them if they work together?
    (a) $3\frac{1}{2}$ days
    (b) 3 days
    (c) $2\frac{2}{7}$ days
    (d) $1\frac{5}{7}$ days

719. A recipe serves four people and calls for $1\frac{1}{2}$ cups of broth. If you want to serve six people, how much broth do you need?
    (a) 2 cups
    (b) $2\frac{1}{4}$ cups
    (c) $2\frac{1}{3}$ cups
    (d) $2\frac{1}{2}$ cups

720. How much water must be added to one gallon of 8% saline solution to get a 2% saline solution?
    (a) 1 gallon
    (b) 2 gallons
    (c) 3 gallons
    (d) 4 gallons

SET 45

721. A twenty-foot piece of wire is cut into three pieces. The first piece is twice as long as the second piece, and four times as long as the third piece. How long is the second piece of wire?

(a) $11\frac{3}{7}$ feet

(b) $5\frac{5}{7}$ feet

(c) $3\frac{2}{7}$ feet

(d) $2\frac{5}{7}$ feet

722. Raj is four times as old as Ram, who is one-third as old as Ravi. If Ravi is 18, what is the sum of their ages?
(a) 64
(b) 54
(c) 48
(d) 24

723. Leela was $\frac{1}{4}$ as young as Kiran five years ago. If the sum of their ages is 110, how old is Leela?
(a) 20
(b) 25
(c) 65
(d) 85

724. Three coolers of water per game are needed for a baseball team of 25 players. If the roster is expanded to 40 players, how many coolers are needed?
(a) 4
(b) 5
(c) 6
(d) 7

725. A Laddoo recipe calls for $1\frac{1}{2}$ cups of besan in order to make 14 Laddoos. If $2\frac{1}{4}$ cups of besan are used, how many Laddoos can be made?
(a) 18
(b) 21
(c) 24
(d) 27

726. The perimeter of a triangle is 25 inches. If side a is twice side b, which is $\frac{1}{2}$ side c, what is the length of side b?
(a) 5
(b) 8
(c) 10
(d) 15

727. How many gallons of a solution that is 75% antifreeze must be mixed with 4 gallons of a 30% solution to obtain a mixture that is 50% antifreeze?
(a) 2 gallons
(b) 3 gallons
(c) 3.2 gallons
(d) 4 gallons

728. How many kgs of concrete containing 14% cement must be mixed with 150 kgs of concrete with 6% cement to create a mixture that is 11% cement?
(a) 900 kgs
(b) 16.5 kgs
(c) 21 kgs
(d) 250 kgs

1000 MATH PROBLEMS >>> Algebra

729. A shopper can spend no more than Rs 20 per kg on fruits and wants 7 kg of bananas at a cost of 50 paise per kg. How many kgs of rasberries can he buy if raspberries cost Rs. 4 per kgs.
(a) 5.25 kgs
(b) 0.5 kgs
(c) 5.75 kgs
(d) 2 kgs

730. A computer costs Rs. 40,000 to purchase. Renting the same computer requires making Rs 5,000 non-refundable deposit plus monthly payments of Rs 2,500. After how many months will the cost to rent the computer equal the cost to purchase it?
(a) 8 months
(b) 14 months
(c) 16 months
(d) 24 months

731. Giri and Mona make Rs. 1,460 together one week. Giri makes Rs. 20 per hour, and Mona makes Rs. 25 per hour. If Giri worked 10 hours more than Mona for one week, how many hours did Mona work?
(a) 25
(b) 28
(c) 33
(d) 38

732. A fruit vendor must pay Rs. 0.90 per kilo for apples and Rs. 1.49 per for kilo grapes. If she buys 47 kilos of apples and 19 kilos of grapes, how much does the fruits cost in all?
(a) Rs. 157.74
(b) Rs. 70.61
(c) Rs. 59.40
(d) Rs. 42.30

733. How much simple interest is earned on Rs. 767 if it is deposited in a bank account paying an annual interest rate of $7\frac{1}{8}$ percent interest for nine months? (Interest = Principal × Rate × Time, or I = PRT)
(a) Rs. 20.56
(b) Rs. 64.13
(c) Rs. 40.99
(d) Rs. 491.83

734. A neighbour has three dogs. Vicky is half the age of Mony, who is one-third as old as Sanju who is half the neighbour's age, which is 24. How old is Vicky?
(a) 2 years
(b) 4 years
(c) 6 years
(d) 12 years

735. A piggy bank contains Rs. 8.20 in coins. If there are an equal number of quarters, nickels, dimes, and pennies, how many of each denomination are there?
(a) 10
(b) 20
(c) 30
(d) 40

736. How many kgs of nuts costing Rs 7 per kilo must be mixed with 6 kgs of nuts costing Rs. 3 per kilo to yield a mixture costing Rs. 4 per kilo?
(a) 2 kilos
(b) 2.5 kilos
(c) 3.5 kilos
(d) 6 kilos

SET 47

737. If Savita deposits Rs. 385 today into a savings account paying 4.85% simple interest annually, how much interest will accrue in one year? (Simple interest earned equals Principal × Rate × Time, or I = PRT.)
    (a) Rs. 1.86
    (b) Rs. 18.67
    (c) Rs. 186.73
    (d) Rs. 1867.245

738. Veena took a trip to the lake. If she drove steadily for 5 hours travelling 220 miles, what was her average speed for the trip?
    (a) 44 mph
    (b) 55 mph
    (c) 60 mph
    (d) 66 mph

739. A problem that begins: "Arvind is $\frac{1}{3}$ as old as his uncle was 20 years ago" is probably going to ask you to
    (a) find the inequality
    (b) subtract Arvind's age from his uncle's
    (c) solve for x
    (d) substitute Arvind's age for his uncle's

740. A dosage of a certain medication is 12 cc per 100 kgs. What is the dosage for a patient who weighs 175 kgs?
    (a) 15 cc
    (b) 18 cc
    (c) 21 cc
    (d) 24 cc

741. A mini van costs Rs. 31,600 to purchase. Leasing the same mini-van requires making Rs. 1,000 deposit plus monthly payments of Rs. 340. After how many years will the cost to lease the minivan equal the cost to purchase it?
    (a) 7.5 years
    (b) 7.75 years
    (c) 8 years
    (d) 90 years

742. How many kgs of solder with 65% tin and 35% lead must be combined with another type of solder with 25% tin and 75% lead to make 80 kgs of solder that is 50% tin and 50% lead?
    (a) 50 kgs
    (b) 40 kgs
    (c) 52 kgs
    (d) 25 kgs

743. If jogging for one mile uses 150 calories and brisk walking for one mile uses 100 calories, a jogger has to go how many times as far as a walker to use the same number of calories?
    (a) $\frac{1}{2}$
    (b) $\frac{2}{3}$
    (c) $\frac{3}{2}$
    (d) 2

744. How much water must be added to 1 litre of a 5% saline solution to get a 2% saline solution?
    (a) 1 L
    (b) 1.5 L
    (c) 2 L
    (d) 2.5 L

1000 MATH PROBLEMS >>> Algebra

745. Akshay will be twice Sunil's age in 3 years when Sunil will be 40. How many years old is Akshay now?
(a) 20
(b) 80
(c) 77
(d) 37

746. A patient's hospice stay cost $\frac{1}{4}$ as his visit to the emergency room. His home nursing cost twice as much as his hospice stay. If his total health care bill was Rs. 140,000, how much did his home nursing cost?
(a) Rs. 10,000
(b) Rs. 20,000
(c) Rs. 40,000
(d) Rs. 80,000

747. Water is coming into a tank three times as fast as it is going out. After one hour, the tank contains 11,400 gallons of water. How fast is the water coming in?
(a) 3,800 gallons/hour
(b) 5,700 gallons/hour
(c) 11,400 gallons/hour
(d) 17,100 gallons/hour

748. Ram is half as old as Sam, who is three times as old as Tony. The sum of their ages is 55. How old is Ram?
(a) 5
(b) 8
(c) 10
(d) 15

749. A man fishing on a riverbank sees a boat pass by. The man estimates the boat is travelling 20 mph. If his estimate is correct, how many minutes will it be before he sees the boat disappear around a bend in the river $\frac{1}{2}$ mile away?
(a) 14 minutes
(b) 1.4 minutes
(c) 2.4 minutes
(d) 1.5 minutes

750. After three days, some hikers discover that they have used $\frac{2}{5}$ of their supplies. At this rate, how many more days can they go forward before they have to turn around?
(a) 0.75 days
(b) 1.5 days
(c) 3.75 days
(d) 4.5 days

751. Kiran was half the age of her mother 20 years ago. Kiran is 40. How old is Kiran's mother?
(a) 50
(b) 60
(c) 70
(d) 80

752. Dr. Khanna charges Rs. 36.00 for an office visit, which is $\frac{3}{4}$ of what Dr. Shah charges. How much does Dr. Shah charge?
(a) Rs. 48.00
(b) Rs. 27.00
(c) Rs. 38.00
(d) Rs. 57.00

SET 48

753. Anju was $\frac{1}{3}$ as young as her grandfather 15 years ago. If the sum of their ages is 110, how old is Anju's grandfather?
(a) 80
(b) 75
(c) 65
(d) 60

754. Five oranges, when removed from a basket containing three more than seven times as many oranges, leaves how many in the basket?
(a) 21
(b) 28
(c) 33
(d) 38

755. How many ounces of candy costing Rs. 1 per ounce must be mixed with 6 ounces of candy costing 70 paise per ounce to yield a mixture costing 80 paise per ounce?
(a) $\frac{6}{7}$ ounce
(b) 3 ounces
(c) 9 ounces
(d) $20\frac{7}{10}$ ounces

756. Dinoo can wash and wax a car in 4 hours. Giri can do the same job in 3 hours. If both work together how long will it take to wash and wax the car?
(a) 1.7 hours
(b) 2.3 hours
(c) 3 hours
(d) 3.3 hours

757. How long will it take Shaila to walk to a store five miles away if she walks at a steady pace of 3 mph?
(a) 0.60 hours
(b) 1.67 hours
(c) 3.33 hours
(d) 15 hours

758. Sam was 10 minutes early for class. Dinesh came in four minutes after Mona, who was half as early as Sam. How many minutes early was Dinesh?
(a) 1 minute
(b) 2 minutes
(c) 2.5 minutes
(d) 6 minutes

759. Vikram loaned Ravi Rs 45 expecting him to repay Rs 50 in one month. What is the amount of annual simple interest on this loan? (R = the annual rate of simple interest. Interest = Principal × Rate × Time, or I = PRT.)
(a) 5%
(b) 33%
(c) 60%
(d) 133%

760. 12 people entered a room. Three more than two-thirds of these people then left. How many people remain in the room?
(a) 0
(b) 1
(c) 2
(d) 7

761. A 24-inch-tall picture is 20% as tall as the ceiling is high. How high is the ceiling?
(a) 4.8 feet
(b) 10 feet
(c) 12 feet
(d) 120 feet

1000 MATH PROBLEMS >>> Algebra

762. Mamta is building a garden shed. When she helped her neighbour build an identical shed it took them both 22 hours to complete the job. If it would have taken her neighbour, working alone, 38 hours to build the shed, how long will it take Mamta, working alone, to build her shed?
   (a) 33.75 hours
   (b) 41.00 hours
   (c) 41.25 hours
   (d) 52.25 hours

763. If nine candles are blown out on a birthday cake containing seven times as many candles altogether, how many candles are there in all?
   (a) 2
   (b) 16
   (c) 63
   (d) 72

764. A bicyclist passes a farmhouse at 3:14 p.m. At 3:56 the bicyclist passes a second farm house. If the bicyclist is travelling at a uniform rate of 12 mph, how far apart are the farmhouses?
   (a) 1.2 miles
   (b) 3.6 miles
   (c) 8.4 miles
   (d) 17.1 miles

765. A snail starts across a road at a constant pace of four inches per minute. A car is approaching at 45 miles per hour. If the car is 15 miles away how many minutes will it take for the snail to cross the 19.8 foot wide road?
   (a) 1 minute
   (b) 30 minutes
   (c) 49 minutes
   (d) 60 minutes

766. Roma finds the average of her three most recent golf scores by using the following expression, where a, b and c are the three scores: $\frac{a+b+c}{3} \times 100$. Which of the following would also determine the average of her scores?
   (a) $(\frac{a}{3} + \frac{b}{3} + \frac{c}{3}) \times 100$
   (b) $\frac{\frac{a+b+c}{3}}{100}$
   (c) $\frac{(a+b+c) \times 3}{100}$
   (d) $\frac{a+b+c}{3} + 100$

767. Seven gallons are removed from a full 20-gallon tank containing a 15% saline solution. Enough pure water is added to refill the tank. What is the percent saline concentration of the resulting mixture?
   (a) 9.75%
   (b) 90%
   (c) 46.33%
   (d) 65%

768. While driving home from work, Rosy runs over a nail causing a tyre to start leaking. She estimates that her tyre is leaking 1 psi every 20 seconds. Assuming that her tyre leaks at a constant rate and her initial tyre pressure was 36 psi, how long will it take her tyre to completely deflate?
   (a) 1.8 minutes
   (b) 3.6 minutes
   (c) 12 minutes
   (d) 18 minutes

1000 MATH PROBLEMS >>> Alegbra

SET 49

769. Tony needs to work five-sixths of a year to pay off his car loan. If Tony begins working on March 1st, at the end of what month will he first be able to pay off his loan?

(a) June
(b) August
(c) October
(d) December

770. A painter starts with 5 gallons of a paint mixture containing 3% paint thinner. She adds enough paint thinner to make a mixture that is 9% thinner. How many gallons are there in this final mixture?

(a) 0.33 gallon
(b) 0.45 gallon
(c) 5.33 gallons
(d) 5.45 gallons

771. Nicky ate 6 pieces of chicken out of a box holding 10 times what Nicky ate. How many pieces of chicken were there originally?

(a) 40
(b) 50
(c) 60
(d) 70

772. Some birds are sitting in an oak tree. Ten more birds land. More birds arrive until there are a total of four times as many birds as the oak tree had after the ten landed. A nearby maple tree has sixteen fewer than twelve times as many birds as the oak tree had after the ten landed. If both trees now have the same number of birds, how many birds were originally on the oak tree before the first 10 landed?

(a) 4
(b) 7
(c) 16
(d) 24

773. A jar of coins totaling Rs. 4.58 contains 13 quarters and 5 nickels. There are twice as many pennies as there are dimes. How many dimes are there?

(a) 5
(b) 9
(c) 18
(d) 36

774. The frontage of a vacant lot is 130 feet, which is one-half as wide as the back of the lot. Both sides of the lot are $1\frac{2}{3}$ as long as the back. What is the perimeter of (*i.e.*, total distance around) the lot?

(a) 125 feet
(b) 412 feet
(c) 823 feet
(d) 1,256 feet

1000 MATH PROBLEMS >>> Algebra

775. A hiker walks from his car to a distant lake and back again. He walks on smooth terrain for 2 hours until he reaches a 5-mile-long, rocky trail. His pace along the 5-mile-long trail is 2 mph. If he walks steadily with no stops, how long will it take the hiker to complete the entire trip from his car to the lake and back again?
   (a) 4.5 hours
   (b) 9 hours
   (c) 12 hours
   (d) 16 hours

776. The total pressure of a mix of gases in a container is equal to the sum of the partial pressures of each of the gases in the mixture. A mixture contains nitrogen, oxygen, and argon, and the pressure of nitrogen is twice the pressure of oxygen, which is three times the pressure of argon. If the partial pressure of nitrogen is 4 kg per square inch (psi), what is the total pressure?
   (a) 24 psi
   (b) 16 psi
   (c) $14\frac{2}{3}$ psi
   (d) $6\frac{2}{3}$ psi

777. Suppose the amount of radiation that could be received from a microwave oven varies inversely as the square of the distance from it. How many feet away must you stand to reduce your potential radiation exposure to $\frac{1}{16}$ the amount you could receive standing 1 foot away?
   (a) 16 feet
   (b) 4 feet
   (c) 32 feet
   (d) 8 feet

778. After a shopping trip Giri had left 19% of the money he started with. He then bought a cup of coffee which cost Rs. 1.29. If Giri started out with Rs. 50, how much money does he have now?
   (a) Rs. 4.07
   (b) Rs. 8.21
   (c) Rs. 9.50
   (d) Rs. 39.20

779. Seven coworkers are having lunch together. The cheque, which they all share equally, amounts to Rs. 48.72. In addition, five of the seven agree to share equally in the payment of a 15% tip. What total amount must each of these five people contribute?
   (a) Rs. 1.46
   (b) Rs. 6.96
   (c) Rs. 7.49
   (d) Rs. 8.42

780. If Raj must pay an employment agency his first month's salary as a placement fee, how much of his Rs. 28,000 first year's salary will Raj end up with?
   (a) Rs. 2,333
   (b) Rs. 3,360
   (c) Rs. 24,343
   (d) Rs. 25,667

1000 MATH PROBLEMS >>> Alegbra

781. Ravi went hiking in bad weather wearing a backpack filled with 60 kgs of supplies. A couple of miles into his hike, he became tired and discarded supplies equal to $\frac{1}{3}$ of the 60 kgs. A few miles later it started to snow, and he discarded another $\frac{2}{5}$ of the original 60 kgs. How much did Ravi discard altogether during his hike?

(a) 5 kgs
(b) 10 kgs
(c) 20 kgs
(d) 44 kgs

782. If a school buys three computers at a, b, and c rupees each, and the school gets a discount of 90%, which expression would determine the average price paid by the school?

(a) $\frac{0.9 \times (a+b+c)}{3}$
(b) $\frac{(a+b+c)}{0.9}$
(c) $(a + b + c) \times 0.9$
(d) $\frac{(a+b+c)}{3}$

783. Three apples and twice as many oranges add up to one-half the number of cherries in a fruit basket. How many cherries are there?

(a) 11
(b) 18
(c) 21
(d) 24

784. Pradip forgot to replace his gas cap the last time he filled up his car with gas. The gas is evaporating out of his 14-gallon tank at a constant rate of $\frac{1}{3}$ gallon per day. How much gas does Pradip lose in 1 week?

(a) 2 gallons
(b) $2\frac{1}{3}$ gallons
(c) $4\frac{2}{3}$ gallons
(d) 6 gallons

SET 50

785. Which ratio best expresses the following: five hours is what percent of a day?
   (a) $\frac{5}{100} = \frac{x}{24}$
   (b) $\frac{5}{24} = \frac{24}{x}$
   (c) $\frac{5}{24} = \frac{x}{100}$
   (d) $\frac{x}{100} = \frac{24}{5}$

786. White flour and whole wheat flour are mixed together in a ratio of 5 parts white flour to 1 part whole wheat flour. How many kgs of white flour are in 48 kgs of this mixture?
   (a) 8 kgs
   (b) 9.6 kgs
   (c) 40 kgs
   (d) 42 kgs

787. Hanif can sell 20 glasses of lemonade for 10 paise per glass. If he raises the price to 25 paise per glass, Hanif estimates he can sell 7 glasses. If so, how much more money will Hanif make by charging 25 paise instead of 10 paise per glass?
   (a) Rs. 0.25
   (b) Rs. 0
   (c) Rs. 0.10
   (d) Rs. 0.50

788. Raj lives five miles away from school. Ravi lives $\frac{1}{2}$ as far away from school. Dinesh's distance from school is half way between Raj and Ravi's. How far away from school does Dinesh live?
   (a) 1.25 miles
   (b) 3.75 miles
   (c) 6.25 miles
   (d) 7.5 miles

789. Praveena's annual salary is six times as much as Anjali's, who earns five times more than Beena, who earns Rs. 4,000. If Veena earns one-half as much as Pramila, what is Veena's annual salary?
   (a) Rs. 25,000
   (b) Rs. 60,000
   (c) Rs. 90,000
   (d) Rs. 1,10,000

790. Yogita and Geeta are sisters. When one fourth of Geeta's age is taken away from Yogita's age, the result is twice Geeta's age. If Yogita is 9, how old is Geeta?
   (a) 2.25 years old
   (b) 4 years old
   (c) 4.5 years
   (d) 18 years old

791. How many kgs of chocolates which cost Rs. 5.95 per kg must be mixed with 3 kgs of caramels costing Rs. 2.95 per kg to obtain a mixture that costs Rs. 3.95 per kg?
   (a) 1.5 kgs
   (b) 3 kgs
   (c) 4.5 kgs
   (d) 8 kgs

792. Kiran charges Rs. 7.50 per hour to mow a lawn. Sherly charges 1.5 times as much to do the same job. How much does Sherly charge to mow a lawn?
   (a) Rs. 5.00 per hour
   (b) Rs. 11.25 per hour
   (c) Rs. 10.00 per hour
   (d) Rs. 9.00 per hour

1000 MATH PROBLEMS >>> Alegbra

793. Leela saves at three times the rate Hira does. If it takes Leela $1\frac{1}{2}$ years to save Rs. 1,000, how long will it take Hira to save this amount?
(a) 1 year
(b) 3.5 years
(c) 4.5 years
(d) 6 years

794. A five-gallon tank is filled with a solution of 50% water and 50% alcohol. Half of the tank is drained and 2 gallons of water are added. How much water is in the resulting mixture?
(a) 3.25 gallons
(b) 4.50 gallons
(c) 5.00 gallons
(d) 2.50 gallons

795. Eight kgs of sunflower seeds which cost Rs. 3 per kg are mixed with 18 kgs of millet which cost 50 paise per kg. Find the cost per kg of this mixture.
(a) Rs. 2.50 per kg
(b) Rs. 1.97 per kg
(c) Rs. 1.27 per kg
(d) Rs. 2.02 per kg

796. Fire departments commonly use the following formula to find out how far from a wall to place the base of a ladder: (Length of ladder ÷ 5 feet) + 2 feet = distance from the wall. Using this formula, if the base of a ladder is placed 10 feet from a wall, how tall is the ladder?
(a) 40 feet
(b) 48 feet
(c) 72 feet
(d) 100 feet

797. Vicky has one-third as many toys as Tom who has four times more toys than Viren, who has 6. How many toys does Jona have if she has 5 more than Vicky.
(a) 4 toys
(b) 13 toys
(c) 8 toys
(d) 9 toys

798. Sushma and Jona are sign painters. Sushma can paint a sign in 6 hours while Jona can paint the same sign in 5 hours. If both worked together, how long would it take them to paint this sign?
(a) 2.53 hours
(b) 2.73 hours
(c) 3.00 hours
(d) 3.23 hours

799. Working by himself Krishna can decorate a store window in 3 hours. Mona can do the same job in 2 hours. How long will it take to decorate this store window if both work together?
(a) 1.2 hours
(b) 1.7 hours
(c) 2.5 hours
(d) 2.7 hours

800. A train moving at a constant speed of 90 mph can travel between two cities in 3.25 hours. How far apart are the cities?
(a) 28 miles
(b) 55 miles
(c) 293 miles
(d) 315 miles

1000 MATH PROBLEMS >>> Algebra

SET 51

801. Raj is three times as old as Nakul. Nakul is twice the age of Ram. The sum of their ages is 117 years. How many years old is Nakul?
    (a) 78 years
    (b) 26 years
    (c) 44 years
    (d) 19 years

802. How much simple interest is earned on Rs. 300 deposited for 30 months in a savings account paying $7\frac{3}{4}\%$ simple interest annually?
    (Simple interest equals Principal × Rate × Time, or I = PRT).
    (a) Rs. 5.81
    (b) Rs. 19.76
    (c) Rs. 23.25
    (d) Rs. 58.13

803. James is six times as old as Kiran. In two years, James will be twice as old as Kiran is then. How old is James now?
    (a) 6 months old
    (b) 3 years old
    (c) 6 years old
    (d) 9 years old

804. Mithilesh types three times as fast as Nitin. Together they type 24 pages per hour. If Nitin learns to type as fast as Mithilesh, how much will they be able to type per hour?
    (a) 18 pages
    (b) 30 pages
    (c) 36 pages
    (d) 40 pages

805. 13 percent of a certain school's students are A students, 15 percent are C students, and 20 percent make mostly Ds. If 16 percent of the students are B students, what percent are failing?
    (a) 25%
    (b) 36%
    (c) 48%
    (d) 64%

806. How much money must be deposited today into a Certificate of Deposit, paying $5\frac{3}{8}\%$ per year simple interest, in order to have Rs. 1,000 in one year?
    (a) Rs. 51.00
    (b) Rs. 53.75
    (c) Rs. 946.25
    (d) Rs. 949.00

807. A four kg mixture of raisins and nuts is $\frac{2}{3}$ raisins. How many kgs of nuts are there?
    (a) 1.3 kgs
    (b) 1.6 kgs
    (c) 2.4 kgs
    (d) 2.6 kgs

808. When estimating distances from aerial photographs, the formula is (A × 1) ÷ f = D, where A is the altitude, 1 is the measured length on the photograph, f is the focal length of the camera, and D is the actual distance. If the focal length is 6 inches, the measured length across a lake on a photograph is 3 inches, and the lake is 3000 feet across, at what altitude was the photograph taken?
    (a) 1500 feet
    (b) 3000 feet
    (c) 4500 feet
    (d) 6000 feet

1000 MATH PROBLEMS >>> Alegbra

809. Pran and Chetna, working together, can sew a dress in nine hours. Working alone, Ram can sew the same dress in 15 hours. How long will it take Pran to sew this dress by herself?
(a) 12 hours
(b) 16 hours
(c) 19.25 hours
(d) 22.5 hours

810. A new telephone directory with 596 pages has 114 less than twice as many pages as last year's edition. How many pages were in last year's directory?
(a) 298 pages
(b) 355 pages
(c) 412 pages
(d) 482 pages

811. A vacationing family travels 300 miles on their first day. The second day they travel only $\frac{2}{3}$ as far. But on the third day they are able to travel $\frac{3}{4}$ as many miles as the first two days combined. How many miles have they travelled altogether?
(a) 375 miles
(b) 875 miles
(c) 525 miles
(d) 500 miles

812. A rain barrel contained 4 gallons of water just before a thunderstorm. It rained steadily for 8 hours, filling the barrel at a rate of 6 gallons per day. How many gallons of water did the barrel have after the thunderstorm?
(a) 4 gallons
(b) 6 gallons
(c) 7 gallons
(d) 9 gallons

813. Jenny will be one-fourth Hari's age on her next birthday when Hari will be twice as old as Jenny's sister, Nishi. Nishi will then be 10 years younger than Hari's brother, Raj. If Raj is 28 on Jenny's birthday, how old will Jenny be?
(a) 9 years old
(b) 14 years old
(c) 18 years old
(d) 26 years old

814. Three species of songbirds live in a certain plot of trees, totaling 120 birds. If species A has 3 times as many birds as species B, which has half as many birds as C, how many birds belong to species C?
(a) 20
(b) 30
(c) 40
(d) 60

815. The base of a rectangle is seven times the height. If the perimeter is 32 meters, what is the area?
(a) 28 m^2
(b) 24 m^2
(c) 16 m^2
(d) 14 m^2

816. A recipe for fudge calls for a mixture of 18% cocoa and 82% sugar. The cook has on hand and would like to use a 10 kg mixture of 12% cocoa and 88% sugar. How many kgs of pure cocoa must be added to be able to use this mixture in the recipe?
(a) 0.06 kgs
(b) 0.37 kgs
(c) 0.60 kgs
(d) 0.73 kgs

SECTION

6

GEOMETRY

The 12 sets of basic geometry problems in this section involve lines, angles, triangles, rectangles, squares, and circles. For example, you may be asked to find the area or perimeter of a shape, the length of a line, or the circumference of a circle. In addition, the word problems will illustrate how closely geometry is related to the real world and to everyday life.

1000 MATH PROBLEMS >>> Geometry

SET 52

817. A square is a special case of all of the following geometric figures EXCEPT a
    (a) parallelogram
    (b) rectangle
    (c) rhombus
    (d) trapezoid

818. How many faces does a cube have?
    (a) 4 faces
    (b) 6 faces
    (c) 8 faces
    (d) 12 faces

819. A polygon is a plane figure composed of connected lines. How many connected lines must there be to make a polygon?
    (a) 3 or more
    (b) 4 or more
    (c) 5 or more
    (d) 6 or more

820. An acute angle is
    (a) 180 degrees
    (b) greater than 90 degrees
    (c) 90 degrees
    (d) less than 90 degrees

821. A straight angle is
    (a) exactly 180 degrees
    (b) between 90 and 180 degrees
    (c) 90 degrees
    (d) less than 90 degrees

822. A right angle is
    (a) 180 degrees
    (b) greater than 90 degrees
    (c) exactly 90 degrees
    (d) less than 90 degrees

823. Which of the following statements is true?
    (a) parallel lines intersect at right angle.
    (b) parallel lines never intersect.
    (c) perpendicular lines never intersect.
    (d) intersecting lines have two points in common.

824. A triangle has two congruent sides, and the measure of one angle is 40 degrees. Which of the following types of triangles is it?
    (a) isosceles
    (b) equilateral
    (c) right
    (d) scalene

825. A triangle has one 30-degree angle and one 60-degree angle. Which of the following types of triangles is it?
    (a) isosceles
    (b) equilateral
    (c) right
    (d) scalene

826. A triangle has angles of 71 degrees and 62 degrees. Which of the following best describes the triangle?
    (a) acute scalene
    (b) obtuse scalene
    (c) acute isosceles
    (d) obtuse isosceles

827. Which of the following does NOT have parallel line segments?
    (a) rhombus
    (b) a square
    (c) a trapezoid
    (d) a rectangle

1000 MATH PROBLEMS >>> Geometry

828. In a triangle, angle A is 70 degrees and angle B is 30 degrees. What is the measure of angle C?
(a) 90 degrees
(b) 70 degrees
(c) 80 degrees
(d) 100 degrees

829. What is a quadrilateral with two parallel sides and an angle of 54 degrees?
(a) triangle
(b) rectangle
(c) square
(d) parallelogram

830. If pentagon ABCDE is similar to pentagon FGHIJ, and AB = 10, CD = 5, and FG = 30, What is IH?
(a) $\frac{5}{3}$
(b) 5
(c) 15
(d) 30

831. What is the greatest area possible enclosed by a quadrilateral with a perimeter of 24 feet?
(a) 6 square feet
(b) 24 square feet
(c) 36 square feet
(d) 48 square feet

832. What is the difference in area between a square with a base of 4 feet and a circle with a diameter of 4 feet?
(a) $16 - 2\pi$ square feet
(b) $16 - 4\pi$ square feet
(c) $8\pi - 16$ square feet
(d) $16\pi - 16$ square feet

SET 53

833. What is the difference in perimeter between a square with a base of 4 feet and a circle with a diameter of 4 feet?
    (a) $8 - 2\pi$ square feet
    (b) $16 - 2\pi$ square feet
    (c) $16 - 4\pi$ square feet
    (d) $16 - 8\pi$ square feet

834. A rectangle's topmost side is 3 times that of the leftmost side. If the leftmost side is A inches long, what is the area of the rectangle?
    (a) $3A$
    (b) $6A$
    (c) $3A^2$
    (d) $6A^2$

835. What is the perimeter of the triangle shown below?

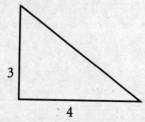

    (a) 12
    (b) 9
    (c) 8
    (d) 7

836. What is the perimeter of the polygon shown below?

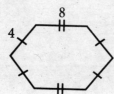

    (a) 12
    (b) 16
    (c) 24
    (d) 32

837. What is the perimeter of the following figure?

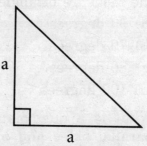

    (a) $a^2 \div 2$
    (b) $2a + 2a^2$
    (c) $2a + \sqrt{2a^2}$
    (d) $4a$

838. The perimeter of a rectangle is 148 feet. Its two longest sides add up to 86 feet. What is the length of each of its two shortest sides?
    (a) 31 feet
    (b) 42 feet
    (c) 62 feet
    (d) 72 feet

839. How many feet of ribbon will a theatrical company need to tie off a performance area that is 34 feet long and 20 feet wide?
    (a) 54
    (b) 68
    (c) 88
    (d) 108

840. What is the outer perimeter of the doorway shown below?

(a) 12
(b) 24
(c) 20 + 2π
(d) 24 + 2π

841. What is the perimeter of the triangle below?

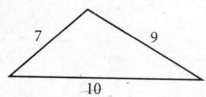

(a) 90
(b) 70
(c) 26
(d) 19

842. What is the perimeter of the parallelogram shown below?

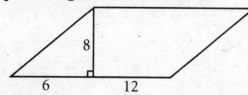

(a) 26
(b) 32
(c) 48
(d) 56

843. What is the perimeter of the shaded area, if the shape is a quarter circle with a radius of 8?

(a) 2π
(b) 4π
(c) 2π + 16
(d) 4π + 16

844. What is the perimeter of the triangle below?

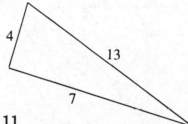

(a) 11
(b) 13
(c) 24
(d) 28

845. What is the perimeter of the polygon shown below?

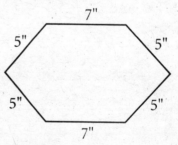

(a) 20 inches
(b) 27 inches
(c) 30 inches
(d) 34 inches

846. What is the perimeter of the rectangle below?

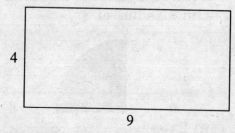

(a) 13
(b) 22
(c) 26
(d) 36

847. What lines are parallel in the following diagram?

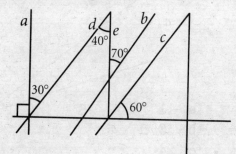

(a) d and b
(b) a and e
(c) e and d
(d) d and c

848. What is the measure of angle F in the following diagram?

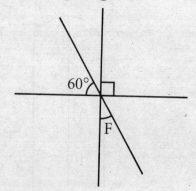

(a) 15 degrees
(b) 45 degrees
(c) 30 degrees
(d) 90 degrees

1000 MATH PROBLEMS >>> Geometry

SET 54

849. A rotating door, pictured below, has 4 sections, labelled a, b, c, and d. If section a is making a 45-degree angle with wall 1, what angle is section c making with wall 2?

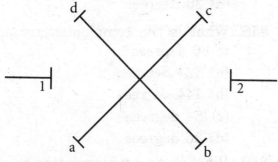

(a) 20 degrees
(b) 45 degrees
(c) 55 degrees
(d) 135 degrees

850. For a driving test, students are to drive around two parallel rows of cones, as shown in the diagram. As the car approaches cone A, it makes a 30-degree angle with the row. What angle must the car turn to meet the opposite row at a 55-degree angle?

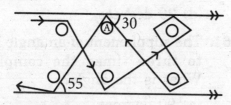

(a) 55 degrees
(b) 60 degrees
(c) 85 degrees
(d) 95 degrees

851. Which side of the right triangle shown below is the shortest, if angle ACB is 46 degrees?

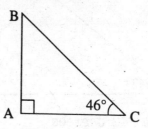

(a) AB
(b) AC
(c) BC
(d) All sides are equal

852. Which side of the triangle shown below is the shortest, if angles BAC and ABC are 60 degrees?

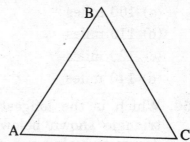

(a) AB
(b) AC
(c) BC
(d) All sides are equal

853. Which of the angles are congruent? (Lines l and m are parallel).

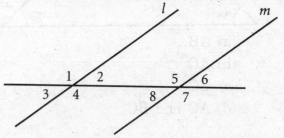

(a) 1 and 2
(b) 1 and 7
(c) 2 and 7
(d) 4 and 8

1000 MATH PROBLEMS >>> Geometry

854. What is the measure of angle ABC?

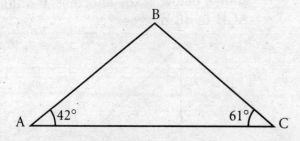

(a) 42 degrees
(b) 61 degrees
(c) 77 degrees
(d) 103 degrees

855. Plattville is 80 miles west and 60 miles north of Quincy. How long is a direct route from Plattville to Quincy?

(a) 100 miles
(b) 110 miles
(c) 120 miles
(d) 140 miles

856. Which is the longest side of the triangle shown below? (Note: not drawn to scale.)

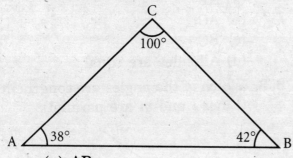

(a) AB
(b) AC
(c) BC
(d) AC and BC

857. An angle has twice the measure of its complement. What is its measure?

(a) 120 degrees
(b) 60 degrees
(c) 45 degrees
(d) 30 degrees

858. What is the complementary angle to 36 degrees?

(a) 324 degrees
(b) 144 degrees
(c) 54 degrees
(d) 36 degrees

859. What is the supplementary angle to 137 degrees?

(a) 137 degrees
(b) 133 degrees
(c) 47 degrees
(d) 43 degrees

860. Three times an angle is equal to two times its complement. What is the angle?

(a) 180 degrees
(b) 120 degrees
(c) 36 degrees
(d) 30 degrees

861. The supplement of an angle is equal to three times the complement. What is the angle?

(a) 90 degrees
(b) 60 degrees
(c) 45 degrees
(d) 30 degrees

1000 MATH PROBLEMS >>> Geometry

862. Which of these angle measures form a right triangle?

(a) 40 degrees, 40 degrees, 100 degrees
(b) 20 degrees, 30 degrees, 130 degrees
(c) 40 degrees, 40 degrees, 40 degrees
(d) 40 degrees, 50 degrees, 90 degrees

863. What is the measure of angle B in the diagram below?

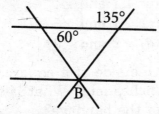

(a) 45 degrees
(b) 60 degrees
(c) 75 degrees
(d) 130 degrees

864. If the figure below is a regular decagon with a centre at Q, what is the measure of the indicated angle?

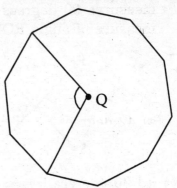

(a) 45 degrees
(b) 80 degrees
(c) 90 degrees
(d) 108 degrees

SET 55

865. In the figure below, angle POS measures 90 degrees. What is the measure of angle ROQ?

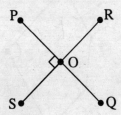

(a) 30 degrees
(b) 45 degrees
(c) 90 degrees
(d) 180 degrees

866. Triangle RST and MNO are similar. What is the length of line segment MO?

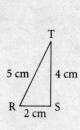 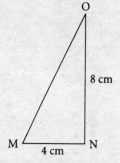

(a) 5 cm
(b) 10 cm
(c) 20 cm
(d) 32 cm

867. In the diagram, lines *a*, *b* and *c* intersect at point O. Which of the following are NOT adjacent angles?

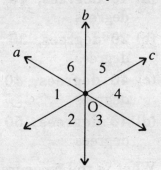

(a) $\angle 1$ and $\angle 6$
(b) $\angle 1$ and $\angle 4$
(c) $\angle 4$ and $\angle 5$
(d) $\angle 2$ and $\angle 3$

868. One side of a square bandage is 4 inches long. What is the perimeter of the bandage?

(a) 4 inches
(b) 8 inches
(c) 12 inches
(d) 16 inches

869. Find the perimeter of the shape below.

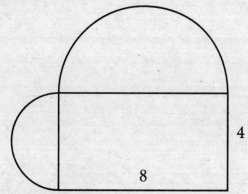

(a) $12 + 10\pi$
(b) $32 + 12\pi$
(c) $32 + 10\pi$
(d) $12 + 12\pi$

1000 MATH PROBLEMS >>> Geometry

870. A triangle has sides that are consecutive even integers. The perimeter of the triangle is 24 inches. What is the length of the shortest side?

(a) 10 inches
(b) 8 inches
(c) 6 inches
(d) 4 inches

871. If the area of a circle is 16π square inches, what is the perimeter?

(a) 2π inches
(b) 4π inches
(c) 8π inches
(d) 16π inches

872. What is the length of one side of a square rug whose perimeter is 60 feet?

(a) 14.5 feet
(b) 15 feet
(c) 15.5 feet
(d) 16 feet

873. What is the perimeter of a pentagon with three sides of 3 inches, and the remaining sides 5 inches long?

(a) 19 inches
(b) 9 inches
(c) 14 inches
(d) 12 inches

874. What is the perimeter of the figure shown?

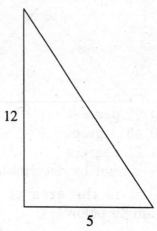

(a) 17
(b) 20
(c) 30
(d) 40

875. If the two triangle in the diagram are similar, with ∠A equal to ∠D, what is the perimeter of triangle DEF?

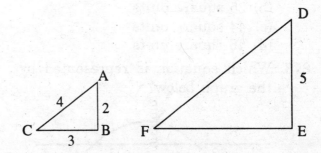

(a) 12
(b) 21
(c) 22.5
(d) 24.75

1000 MATH PROBLEMS >>> Geometry

876. What is the measure of angle C in the following triangle?

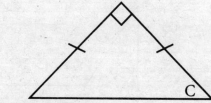

(a) 90 degrees
(b) 45 degrees
(c) 25 degrees
(d) cannot be determined

877. What is the area of the shaded triangle below?

(a) 20 square units
(b) 25 square units
(c) 44 square units
(d) 46 square units

878. Which equation is represented by the graph below?

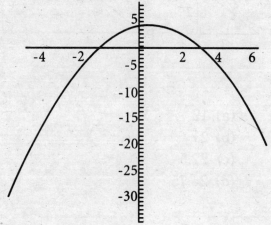

(a) $y = x^2 + 2x = 4$
(b) $y = -x^2 + 2x - 4$
(c) $y = x^2 + 2x + 4$
(d) $y = x^2 + 2x - 4$

879. Which side is the longest if triangle A is similar to triangle B?

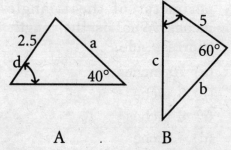

(a) a
(b) b
(c) c
(d) d

880. In the diagram below, if angle 1 is 30° and angle 2 is a right angle, what is the measure of angle 5?

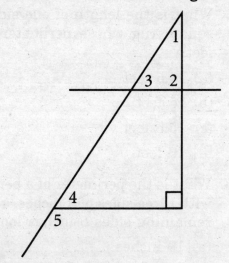

(a) 30 degrees
(b) 60 degrees
(c) 120 degrees
(d) 140 degrees

SET 56

881. What is angle *a* in the following diagram?

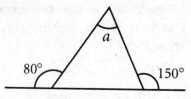

(a) 70 degrees
(b) 60 degrees
(c) 50 degrees
(d) 40 degrees

882. What is the measure of angle ABC if ABCD is a parallelogram, and the measure of angle BAD is 88 degrees?

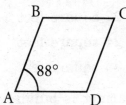

(a) 88 degrees
(b) 90 degrees
(c) 92 degrees
(d) 102 degrees

883. One base angle of an isosceles triangle is 70 degrees. What is the vertex angle?

(a) 130 degrees
(b) 90 degrees
(c) 70 degrees
(d) 40 degrees

884. A circular fan is encased in a square guard. If one side of the guard is 12 inches, at what blade circumference will the fan just hit the guard?

(a) 6 inches
(b) 12 inches
(c) 6π inches
(d) 12π inches

885. If the circumference of a circle is half the area, what is the radius of the circle?

(a) $\frac{1}{2}$
(b) 2
(c) 4
(d) 8

886. What is the circumference of a circle with a diameter of 5 inches?

(a) 2.5π inches
(b) 5π inches
(c) 6.25π inches
(d) 25π inches

887. What is the area of the following diagram?

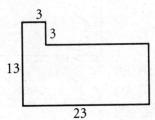

(a) 239
(b) 259
(c) 299
(d) 306

888. What is the perimeter of the following rectangle?

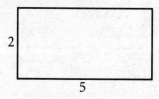

(a) 7
(b) 10
(c) 14
(d) 70

889. What is the volume of a pyramid that has a rectangular base 5 feet by 3 feet and a height of 8 feet? Use $V = \frac{1}{3}$ (area of base)(height).

(a) 16 cubic feet
(b) 30 cubic feet
(c) 40 cubic feet
(d) 80 cubic feet

890. Giri is making a box. He starts with a 10-by-7 rectangle, then cuts 2-by-2 squares out of each corner. To finish, he folds each side up to make the box. What is the box's volume?

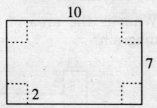

(a) 36
(b) 70
(c) 72
(d) 140

891. In order to protect her new Maruti Zen, Mallika needs to build a new garage. The concrete floor needs to be 64.125 square feet and is $9\frac{1}{2}$ feet long. How wide does it need to be?

(a) 7.25 feet
(b) 5.5 feet
(c) 6.75 feet
(d) 8.25 feet

892. All of the rooms on the top floor of a government building are rectangular, with 8-foot ceilings. One room is 9 feet wide by 11 feet long. What is the combined area of the four walls, including doors and windows?

(a) 99 square feet
(b) 160 square feet
(c) 320 square feet
(d) 72 square feet

893. A rectangular tumbling mat for a gym class is 5 feet wide and 7 feet long. What is the area of the mat?

(a) 12 square feet
(b) 22 square feet
(c) 24 square feet
(d) 35 square feet

894. A farmer is building a rectangular pen on the side of his barn, which is 100 feet long. He has 300 feet of fence and is using the side of the barn as the fourth side of the fence. What will be the area of the pen?

(a) 10,000 square feet
(b) 20,000 square feet
(c) 30,000 square feet
(d) 90,000 square feet

895. Lina wants to wallpaper a room. It has one window that measures 3 feet by 4 feet, and one door that measures 3 feet by 7 feet. The room is 12 feet by 12 feet, and is 10 feet tall. If only the walls are to be covered, and rolls of wallpaper are 100 square feet, what is the minimum number of rolls that she will need?

(a) 4 rolls
(b) 5 rolls
(c) 6 rolls
(d) 7 rolls

896. What is the area of the triangle shown below?

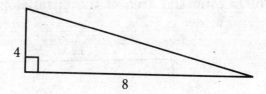

(a) 6
(b) 12
(c) 16
(d) 32

SET 57

897. Find the area of the parallelogram below.

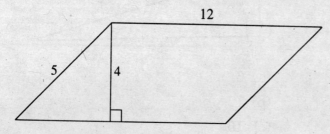

(a) 48
(b) 68
(c) 72
(d) 240

898. What is the area of the rectangle?

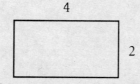

(a) 6
(b) 8
(c) 12
(d) 16

899. A hospital waiting room is 8 feet wide and 10 feet long. What is the area of the waiting room?
(a) 18 square feet
(b) 40 square feet
(c) 60 square feet
(d) 80 square feet

900. The length of a rectangle is equal to 4 inches more than twice the width. Three times the length plus two times the width is equal to 28 inches. What is the area of the rectangle?
(a) 8 square inches
(b) 16 square inches
(c) 24 square inches
(d) 28 square inches

901. A rectangular box has a square base with an area of 9 square feet. If the volume of the box is 36 cubic feet, what is the length of the longest object that can fit in the box?
(a) 3 feet
(b) 5 feet
(c) 5.8 feet
(d) 17 feet

902. Dinesh is buying land on which he plans to build a cabin. He wants 200 feet in road frontage and a lot 500 feet deep. If the asking price is Rs. 9,000 an acre for the land, how much will Dinesh pay for his lot?
(a) Rs. 10,000
(b) Rs. 20,700
(c) Rs. 22,956
(d) Rs. 24,104

903. In the following diagram, a circle of area 100π square inches is inscribed in a square. What is the length of side AB?

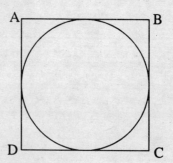

(a) 10 inches
(b) 20 inches
(c) 100 inches
(d) 400 inches

904. Geeta is making a quilt. She wants a quilt that is 30 square feet. She has collected fabric squares that are 6 inches by 6 inches. How many squares will she need?

(a) 60 squares
(b) 90 squares
(c) 100 squares
(d) 120 squares

905. What is the area of the following rectangle?

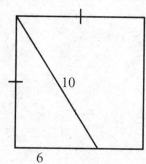

(a) 30
(b) 60
(c) 64
(d) 150

906. What is the value of X in the figure below?

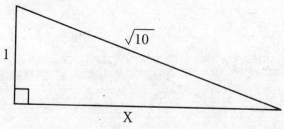

(a) 3
(b) 4
(c) 5
(d) 9

907. What is the area of the shaded figure inside the rectangle?

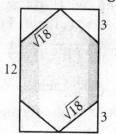

(a) 18
(b) 54
(c) 60
(d) 72

908. Arun has a canvas frame that is 25 inches long and 18 inches wide. He buys a canvas that is 3 inches longer on each side. What is the area of the canvas?

(a) 450 square inches
(b) 744 square inches
(c) 588 square inches
(d) 872 square inches

909. Prisoner Johny escaped a short time ago. On foot, he has not gone far, and is believed to be within a 3-mile radius of the prison. What is the approximate area, in square miles, of the area in which the prisoner is hiding?

(a) 28 square miles
(b) 30 square miles
(c) 9 square miles
(d) 10 square miles

1000 MATH PROBLEMS >>> Geometry

910. Mira threw her book at the yowling cat. Fortunately, it fell short, and landed with its spine up, in an isosceles triangle. The height of the book triangle is $7\frac{1}{2}$ inches and the base is $5\frac{1}{2}$ inches. What is the area of the triangle formed by the tossed book?

(a) 41.25 square inches
(b) 20.6 square inches
(c) 32.5 square inches
(d) 36.75 square inches

911. How many six-inch square tiles are needed to tile the floor in a room that is 12 feet by 15 feet?

(a) 180 tiles
(b) 720 tiles
(c) 540 tiles
(d) 360 tiles

912. Hena's backyard swimming pool is round, with an area of 113 feet. What is the diameter of the pool?

(a) 9 feet
(b) 10 feet
(c) 11 feet
(d) 12 feet

SET 58

913. A two-storey house is 20 feet high. The sides of the house are 28 feet long; the front and back are each 16 feet long. A gallon of paint will cover 440 square feet. How many gallons are needed to paint the whole house?

(a) 3 gallons
(b) 4 gallons
(c) 5 gallons
(d) 6 gallons

914. Jona has a square sandbox; each side is 6 feet-long. What is the area of the sandbox?

(a) 12 square feet
(b) 24 square feet
(c) 36 square feet
(d) 48 square feet

915. A Persian rug is 6.5 feet by 8 feet. What is the area of the rug?

(a) 48 square feet
(b) 52 square feet
(c) 29 square feet
(d) 14.5 square feet

916. The kitchen in Khanna's old house contains a circular stone inset with a radius of 3 feet in the middle of the floor. The room is square; each wall is 11.5 feet long. Khanna wants to re-tile the kitchen, all except for the stone inset. What is the approximate square footage of the area Khanna needs to buy tiles for?

(a) 76 square feet
(b) 28 square feet
(c) 132 square feet
(d) 104 square feet

917. What is the area of a triangular lot that is 75 feet at the sidewalk and 167 feet long?

(a) 6262.5 square feet
(b) 12525 square feet
(c) 4843.5 square feet
(d) 16775 square feet

918. Kavita is running for Student Council President. The rules restrict the candidates to four 2-foot by 3-foot posters. Kavita has dozens of 4-inch by 6-inch pictures of herself, taken by her boyfriend. What's the maximum number of pictures she will be able to use on the 4 posters?

(a) 144 pictures
(b) 130 pictures
(c) 125 pictures
(d) 111 pictures

919. Raj has a picture that is 25 inches by 19 inches and a frame that is 30 inches by 22 inches. He will make up the difference with a mat that will fit between the picture and the frame. What is the area of the mat he will need?

(a) 660 square inches
(b) 475 square inches
(c) 250 square inches
(d) 185 square inches

920. Reema must replace a sail on her boat. The triangular sail is 30 feet high and 20 feet at the base. What is the area of the sail?

(a) 600 square feet
(b) 500 square feet
(c) 400 square feet
(d) 300 square feet

1000 MATH PROBLEMS >>> Geometry

921. Nakul needs 45 square yards of fabric for curtains on two windows. If the fabric is 54 inches wide, how many yards should he buy?
   (a) 10 yards
   (b) 15 yards
   (c) 20 yards
   (d) 30 yards

922. The ceiling in a rectangular room is 8 feet high. The room is 15 feet by 12 feet. Bunny wants to paint the long walls Moss and the short walls Daffodil. The paint can only be purchased in gallon cans, and each gallon will cover 200 square feet. How many cans of each colour will Bunny need?
   (a) 2 cans of Moss & 1 can of Daffodil
   (b) 1 can of Moss & 2 cans of Daffodil
   (c) 2 cans of Moss & 2 cans of Daffodil
   (d) 1 can of Moss & 1 can of Daffodil

923. In Delhi, property taxes for residential areas are Rs. 2.53 per square foot. Mrs. Chawala's house is 30 feet wide and 50 feet long. How much is her tax bill?
   (a) Rs. 3,957
   (b) Rs. 7,953
   (c) Rs. 7,539
   (d) Rs. 3,795

924. A rectangular school hallway is to be tiled with 6-inch-square tiles. The hallway is 72 feet long and 10 feet wide. Lockers along both walls narrow the hallway by 1 foot on each side. How many tiles are needed to cover the hallway floor?
   (a) 16
   (b) 20
   (c) 2,304
   (d) 2,880

925. A parking lot is 50 feet by 100 feet. What is its area, in square feet?
   (a) 150
   (b) 500
   (c) 2,500
   (d) 5,000

926. A waiter folds a napkin in half to form a 5-inch square. What is the area, in square inches, of the unfolded napkin?
   (a) 25
   (b) 30
   (c) 50
   (d) 60

927. Savita has a lamp placed in the centre of her yard. The lamp shines a radius of 10 feet on her yard, which is 20 feet on each side. How much of the yard, in square feet, is NOT lit by the lamp?
   (a) 400π
   (b) 100π
   (c) $40 - 10\pi$
   (d) $400 - 100\pi$

928. In the diagram, a half-circle is laid adjacent to a triangle. What is the total area of the shape, if the radius of the half-circle is 3 and the height of the triangle is 4?

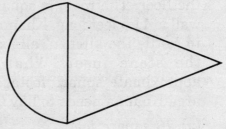

   (a) $6(\pi + 4)$
   (b) $6\pi + 12$
   (c) $6\pi + 24$
   (d) $\frac{9\pi}{2} + 12$

1000 MATH PROBLEMS >>> Geometry

SET 59

929. Mahima is building a mall with 20 stores, each 20 by 35 feet. The mall has one hallway with dimensions of 100 by 20 feet. What is the mall's square footage?
(a) 18,000
(b) 16,000
(c) 6,000
(d) 2,000

930. Prachi is restoring an antique trunk. She is painting the outside pink and the inside white. The trunk is 3 feet long, 16 inches wide, and 2 feet tall. There is a brass latch on the outside of the trunk that takes up an area of 9 square inches. How much paint does she need, in square inches to cover both the inside and outside of the trunk?
(a) 5,368
(b) 7,287
(c) 9,252
(d) 27,639

931. A rotating sprinkler sprays water 10 feet as it rotates in a circle. How much grass will the sprinkler water?
(a) 20π square feet
(b) 100π square feet
(c) 200π square feet
(d) 400π square feet

932. Anjali has a circular swimming pool with a perimeter of 18π feet. She wants to put in a safety rope that spans from one end of the pool to the other. How long should the rope be?
(a) 9 feet
(b) 18 feet
(c) 20 feet
(d) 36 feet

933. Mona is wallpapering a circular room with a 10-foot radius and a height of 8 feet. Ignoring the doors and windows, how much wall paper will she need?
(a) 252π square feet
(b) 70π square feet
(c) 100π square feet
(d) 160π square feet

934. A rectangular area has one side that measures 15 inches and another side one-third as long. What is the area of the rectangle?
(a) 37.5 square inches
(b) 40 square inches
(c) 55 square inches
(d) 75 square inches

935. The base of a triangle is twice the height of the triangle. If the area is 16 square inches, what is the height?
(a) 4 inches
(b) 8 inches
(c) 12 inches
(d) 16 inches

936. What is the length of a rectangular room which has an area of 132 square feet and has a width of 12 feet?
(a) 10
(b) 11
(c) 12
(d) 13

937. What is the area of the following isosceles triangle?

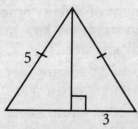

(a) 12 square units
(b) 15 square units
(c) 6 square units
(d) 24 square units

938. What is the area of the parallelogram shown below?

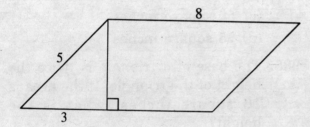

(a) 15
(b) 24
(c) 32
(d) 40

939. What is the area of the following figure?

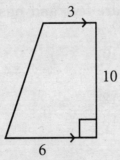

(a) 30 square units
(b) 45 square units
(c) 60 square units
(d) 90 square units

940. What is the measure of the diagonal of the rectangle below?

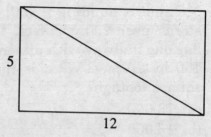

(a) 8.5
(b) 12
(c) 13
(d) 17

941. Find the area of the shaded portion in the figure below.

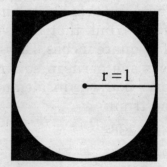

(a) π
(b) π − 1
(c) 2 − π
(d) 4 − π

1000 MATH PROBLEMS >>> Geometry

942. What is the area of the figure shown below?

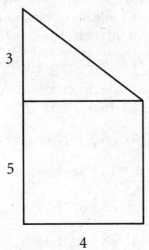

(a) 22
(b) 26
(c) 30
(d) 32

943. What is the difference in area between a square with a base of 4 inches, a slant of 6 inches, and a height of 4 inches; and a parallelogram with a base of 4 inches, a slant of 6 inches, and a height of 4 inches?

(a) 0 square inches
(b) 2 square inches
(c) 4 square inches
(d) 8 square inches

944. The radius of a circle is 13. What is the approximate area of the circle?

(a) 81.64
(b) 530.66
(c) 1,666.27
(d) 169

1000 MATH PROBLEMS >>> Geometry

SET 60

945. Horses are racing on a circular track with a perimeter of 360 feet. Two cameras are in the centre of the track. Camera A is following Horse A and Camera B is following Horse B. If the angle between the two cameras is 40 degrees, how far apart are the 2 horses?

(a) 80π feet
(b) 40π feet
(c) 40 feet
(d) 90 feet

946. Jaya is replacing all the plumbing in her house. The pipe she is using has a diameter of 2.5 inches. What is the circumference of the pipe?

(a) 7.85 inches
(b) 15.7 inches
(c) 5 inches
(d) 6.28 inches

947. A cow is tied to a post with a twenty-feet rope. What is the longest path that the cow can walk?

(a) 20 feet
(b) 40 feet
(c) 20π feet
(d) 40π feet

948. Ganesh is making a doily for the top of a round end table. The diameter of the table is 24 inches. What will be the circumference of the doily if it covers the tabletop?

(a) 48.72 inches
(b) 60.43 inches
(c) 75.36 inches
(d) 82.58 inches

949. Rashmi is using silk to decorate the lids of round tin boxes. She will cut the silk in circles and use fabric glue to attach them. The lids have a radius of $4\frac{1}{2}$ inches. What should be the approximate circumference of the fabric circles?

(a) $14\frac{1}{4}$ inches
(b) $18\frac{1}{2}$ inches
(c) $24\frac{1}{2}$ inches
(d) $28\frac{1}{4}$ inches

950. In order to fell a tree, lumberjacks drive a spike through the centre. If the circumference of a tree is 43.96 feet, what is the minimum length needed to go completely through the tree, passing through the centre?

(a) 7 feet
(b) 14 feet
(c) 21 feet
(d) 28 feet

951. Farmer Rama finds a crop circle in the north forty acres, where he believes an alien spacecraft landed. A reporter from the Daily News measures the circle and finds that its radius is 25 feet. What is the circumference of the crop circle?

(a) 78.5 feet
(b) 157 feet
(c) 208 feet
(d) 357.5 feet

1000 MATH PROBLEMS >>> Geometry

952. Ashi wants to buy a round dining table the same size as her circular rag rug. The circumference of the rug is 18.84 feet. What should be the diameter of the table she buys?

(a) 6 feet
(b) 18 feet
(c) 3.14 feet
(d) 7 feet

953. A pie pan has an area of 78.5 square inches. What is the radius of the pan?

(a) 10 inches
(b) 7 inches
(c) 5 inches
(d) 3 inches

954. Anita is making cookies. She rolls out the dough to a rectangle that is 18 inches by 12 inches. Her circular cookie cutter has a circumference of 9.42 inches. Assuming she reuses the dough scraps, approximately how many cookies can Anita cut of the dough?

(a) 31 cookies
(b) 14 cookies
(c) 12 cookies
(d) 10 cookies

955. Ravi's bike wheel has a diameter of 27 inches. He rides his bike until the wheel turns 100 times. How far did he ride?

(a) $1,500\pi$ inches
(b) $2,700\pi$ inches
(c) $3,200\pi$ inches
(d) $4,800\pi$ inches

956. A three-sided corner lot contains one side that is 50 feet long, one that is 120 feet long, and one that is 150 feet long. What is the perimeter of the lot?

(a) 320 feet
(b) 6000 feet
(c) 640 feet
(d) 7500 feet

957. While walking through the junkyard, Savita and Giri found a stop sign. They decided to take the six-sided sign home and hang it on the bathroom wall. Each of the sides is 12 inches long. What is the perimeter of the stop sign?

(a) 72 inches
(b) 144 inches
(c) 48 inches
(d) 720 inches

958. Naveena wants to hang a garland of silk flowers all around the ceiling of a square room. Each side of the room is 9 feet long; the garlands are only available in 15-foot lengths. How many garlands will she need to buy?

(a) 2 garlands
(b) 3 garlands
(c) 4 garlands
(d) 5 garlands

1000 MATH PROBLEMS >>> Geometry

959. Ram is buying wallpaper for a square room in which the perimeter is 52 feet. How long is each side?
(a) 12 feet
(b) 14 feet
(c) 13 feet
(d) 15 feet

960. Rahul painted a 8-foot by 10-foot canvas floor cloth with a blue border 8 inches wide. What is the perimeter of the unpainted section?
(a) $30\frac{2}{3}$ feet
(b) $15\frac{1}{3}$ feet
(c) $35\frac{1}{3}$ feet
(d) $20\frac{2}{3}$ feet

1000 MATH PROBLEMS >>> Geometry

SET 61

961. The bride's table at a big wedding is 3 feet wide and 48 feet long. How many yards of mauve ribbon will be needed to go around the perimeter of the table?

(a) 102 yards
(b) 51 yards
(c) 34 yards
(d) 26 yards

962. Meena has a fish pond shaped like an equilateral triangle, which is 20 feet on a side. She wants to fill the pond, which is 10 feet deep, with water. How much water will she need?

(a) $(10)(\sqrt{500})$ cubic feet
(b) $(10)(\sqrt{300})$ cubic feet
(c) 200 cubic feet
(d) 2000 cubic feet

963. Aman takes a compass reading and finds that he is 32 degrees north of east. If he was facing the exact opposite direction, what would his compass reading be?

(a) 32 degrees north of west
(b) 58 degrees west of south
(c) 58 degrees north of east
(d) 32 degrees south of east

964. Rani is putting a stairway in her house. If she wants the stairway to make an angle of 20 degrees to the ceiling, what obtuse angle should the stairway make to the floor?

(a) 30 degrees
(b) 40 degrees
(c) 70 degrees
(d) 160 degrees

965. When building a house, the builder should make certain that the walls meet at a 90-degree angle. This is called

(a) a straight angle
(b) an acute angle
(c) an obtuse angle
(d) a right angle

966. Mona cuts a pizza in half in a straight line. She then cuts a line from the centre to the edge, creating a 35-degree angle. What is the supplement of that angle?

(a) 55 degrees
(b) 145 degrees
(c) 35 degrees
(d) 70 degrees

967. In a pinball machine, a ball bounces off a bumper at an angle of 72 degrees and then bounces off another bumper that is parallel to the first. If the angles that the ball comes to and leaves the first bumper are equal, what is the angle that the ball bounces off the second bumper?

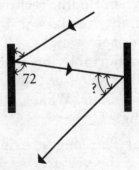

(a) 36 degrees
(b) 48 degrees
(c) 72 degrees
(d) 90 degrees

1000 MATH PROBLEMS >>> Geometry

968. To wash the windows on the second floor of his house, Amar leaned a ladder against the house, creating a 54-degree acute angle with the ground. What is the obtuse angle the ladder made with the ground?

(a) 36 degrees
(b) 126 degrees
(c) 81 degrees
(d) 306 degrees

969. Raj, Geeta, Dinesh, Bunny and their friends are playing stickball on Maple Street. Home base is a trash can lid; 20 feet away is first base, a fire hydrant; 17 feet from the hydrant is second base, an empty beer can; third base, a hub cap, is 20 feet from the can; and back to home base is 17 feet. If the opposite angles thus formed are not 90 degrees, what kind of quadrilateral have the players created?

(a) a rectangle
(b) a square
(c) a triangle
(d) a parallelogram

970. When Sudhir reclines in his lounge chair, he likes it to be at a 135-degree angle. To be really comfortable, he places a pillow in the vertex of the angle formed by the chair. What is the vertex of an angle?

(a) The longest side of an angle is the vertex.
(b) The imaginary third side of an angle is the vertex.
(c) The point of an angle is the vertex.
(d) The formula for figuring the degrees of an angle is the vertex.

971. A hen is standing at the side of a straight section of road. On the same side, a cow is standing 60 feet down the road; on the other side of the road and directly across from the chicken, a bowl of corn sits. The corn is 80 feet from the cow. About how far will the hen have to go to cross the road to eat the corn?

(a) 45 feet
(b) 49 feet
(c) 53 feet
(d) 61 feet

972. Two ships leave from a port. Ship A sails west for 300 miles, and Ship B sails south for 400 miles. How far apart are the ships after their trips?

(a) 300 miles
(b) 400 miles
(c) 500 miles
(d) 900 miles

973. The force of the earthquake caused Amit's house to lean. As a result, the front wall in the living room contains the following angles: corner A is at a 70-degree angle, corner B is at a 110-degree angle, corner C is at a 70-degree angle, and corner D is at a 110-degree angle. What shape has the wall taken on?

(a) parallelogram
(b) square
(c) rectangle
(d) diamond

1000 MATH PROBLEMS >>> Geometry

974. Ram has a round coffee table that has 6 equal triangle-shaped inlays. The smallest angle of each triangle meets in the centre of the table; each triangle touches the edge of the triangle on either side. Three intersecting lines, therefore, create the six identical triangles. What is the angle of each of the points in the centre of the table?

(a) 120 degrees
(b) 90 degrees
(c) 75 degrees
(d) 60 degrees

975. The most ergonomically correct angle between the keyboard and the screen of a laptop computer is 100 degrees. This is known as

(a) an acute angle
(b) a complementary angle
(c) an obtuse angle
(d) a right angle

976. When Raj swam across a river, the current carried him 20 feet downstream. If the total distance that he swam was 40 feet, how wide is the river to the nearest foot?

(a) 30 feet
(b) 32 feet
(c) 35 feet
(d) 50 feet

SET 62

977. While waiting for her test results, Leela paces in a perfectly square room in which each wall is 9 feet long. She walks along one wall to the end, turns 90 degrees and walks along the wall to the end. Then, she crosses back to where she started, creating an isosceles triangle. How long is the third leg—or hypotenuse—of her journey?

(a) 12.7 feet
(b) 11 feet
(c) 16.2 feet
(d) 9 feet

978. Sunder is leaning a ladder against his house so that the top of the ladder makes a 75-degree angle with the house wall. What is the measure of the acute angle where the ladder meets the level ground?

(a) 15 degrees
(b) 25 degrees
(c) 45 degrees
(d) 165 degrees

979. Geeta is building a path in her yard. She wants to bisect her garden, which is 36 feet long. How far from the end of Geeta's garden will the path cross it?

(a) 12 feet
(b) 18 feet
(c) 20 feet
(d) 72 feet

980. Mani, the bike messenger picks up a package at Print Quick, delivers it to Raj's office 22 blocks away, waits for Raj to approve the package, and then hops on his bike and returns the package to Print Quick. If one block is 90 yards, how many yards will Mani bike in total?

(a) 44
(b) 180
(c) 1,980
(d) 3,960

981. Amar is building a brick wall that will measure 10 feet by 16 feet. If the bricks he is using are 3 inches wide and 5 inches long, how many bricks will it take to build the wall?

(a) 128
(b) 160
(c) 1,536
(d) 23,040

982. A power line stretches down a 400-foot country road. A pole is to be put at each end of the road, and 1 in the midpoint of the wire. How far apart is the centre pole from the left-most pole?

(a) 100 feet
(b) 200 feet
(c) 300 feet
(d) 400 feet

1000 MATH PROBLEMS >>> Geometry

983. Kiran walks to school in a straight line; the distance is 7 blocks. The first block is 97 feet long; the second and third blocks are 90 feet long; the fourth, fifth, and sixth blocks are 110 feet long. The seventh block is congruent to the second block. How far does Kiran walk?

(a) 770 feet
(b) 717 feet
(c) 704 feet
(d) 697 feet

984. Main Street and Broadway are parallel to each other; First Avenue and Second Avenue are also parallel to one another and, in addition, are both perpendicular to Main and Broadway. North Boulevard transverses Main Street. What other street(s) does it definitely also cross?

(a) Broadway
(b) Broadway and First Avenue
(c) First Avenue and Second Avenue
(d) Broadway, First Avenue, and Second Avenue

985. Bharat, Raj, Roma, and Dinesh want to share a brownie evenly. Roma cuts one line down the middle and hands the knife to Dinesh. What kind of line should Dinesh cut through the middle of the first line?

(a) a parallel line
(b) a transversal line
(c) a perpendicular line
(d) a skewed line

986. Mangalore, the town lies half in one state and half in another. To make it even more confusing, the straight line that divides the states goes through the houses on Sandusky Street. What is the relationship of the houses at 612, 720 and 814?

(a) the houses are congruent
(b) the houses are collinear
(c) the houses are segmented
(d) the houses are equilateral

987. A triangular piece of wood has a base of 4 inches and a height of 5 inches. What is the area of the piece of wood?

(a) 20 square inches
(b) 9 square inches
(c) 10 square inches
(d) 19 square inches

988. The city requires all pit bulls to be in fenced yards. Your yard is 70 feet by 110 feet. How much chain link fence do you need for your pit bull's yard, excluding a three-foot gate at the sidewalk?

(a) 7,697 feet
(b) 357 feet
(c) 3,570 feet
(d) 360 feet

989. Anand discovers to his horror that the corner of the dining room in his new house is a 105-degree angle. He should sue his contractor for building a room with

(a) an acute angle
(b) a right angle
(c) a straight angle
(d) an obtuse angle

990. A container filled with water is 10 by 10 by 15 inches. Nidhi fills her glass with 60 cubic inches of water from the container. How much water is left in the container?
(a) 1,500 cubic inches
(b) 1,440 cubic inches
(c) 1,000 cubic inches
(d) 60 cubic inches

991. A steel box has a base length of 12 inches and a width of 5 inches. If the box is 10 inches tall, what is the total volume of the box?
(a) 480 cubic inches
(b) 540 cubic inches
(c) 600 cubic inches
(d) 720 cubic inches

992. A water tank is in the form of a right cylinder on top of a hemisphere, both with a radius of 3 feet. If the tank currently has 170 cubic feet of water in it, how high does the water level reach in the cylinder (from the top of the hemisphere)?
(a) 2 feet
(b) 3 feet
(c) 4 feet
(d) 6 feet

1000 MATH PROBLEMS >>> Geometry

SET 63

993. A builder has 27 cubic feet of concrete to pave a sidewalk whose length is 6 times its width. The concrete must be poured 6 inches deep. How long is the sidewalk?

(a) 9 feet
(b) 12 feet
(c) 15 feet
(d) 18 feet

994. Aruna has two containers for water. A rectangular plastic box with a base of 16 square inches, and a cylindrical container with a radius of 2 inches and a height of 11 inches. If the rectangular box is filled with water 9 inches from the bottom, and Aruna pours the water into the cylinder without spilling, which of the following will be true?

(a) the cylinder will overflow
(b) the cylinder will be exactly full.
(c) the cylinder will be filled to an approximate level of 10 inches.
(d) the cylinder will be filled to an approximate level of 8 inches.

995. A cereal box is filled with Snappie cereal at the Snappie production facility. When the box arrives at the supermarket, the cereal has settled and takes up volume of 144 cubic inches. There is an empty space in the box of 32 cubic inches. If the base of the box is 2 by 8 inches, how tall is the box?

(a) 5 inches
(b) 11 inches
(c) 12 inches
(d) 15 inches

996. Sanjeev has a briefcase with dimensions of 2 by $1\frac{1}{2}$ by $\frac{1}{2}$ feet. He wishes to place several notebooks, each 8 by 9 by 1 inch, into the briefcase. What is the largest number of notebooks the briefcase will hold?

(a) 42
(b) 36
(c) 20
(d) 15

997. One cubic foot of copper is to be made into wire with a $\frac{1}{2}$ inch radius. If all of the copper is used to produce the wire, how long will the wire be, in inches?

(a) 2,000
(b) $\frac{4,000}{\pi}$
(c) $\frac{6,912}{\pi}$
(d) $\frac{48,000}{\pi}$

998. Danny wants to know the height of a telephone pole. He measures his shadow, which is 3 feet long, and the pole's shadow, which is 10 feet long. Danny's height is 6 feet. How tall is the pole?

(a) 40 feet
(b) 30 feet
(c) 20 feet
(d) 10 feet

999. Which of the following inequalities is represented by the graph below?

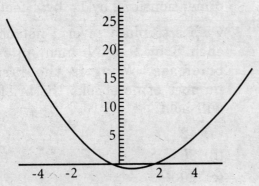

(a) $y \geq x^2 + 2$
(b) $y \geq x^2 - 2$
(c) $y \geq x^2 + 2x$
(d) $y \geq x^2 - 2x$

1000. Which 3 points are both collinear and coplanar?

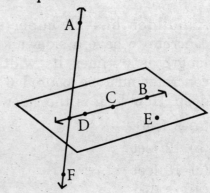

(a) A, B, C
(b) A, C, D
(c) B, C, D
(d) D, E, A

ANSWERS

SECTION 1 : MISCELLANEOUS MATH

SET 1

1. (d) The correct answer to this basic addition problem is 10.
2. (a) The correct answer to this basic subtraction problem is 7.
3. (b) The correct answer to this basic division problem is 4.
4. (d) The correct answer to this simple multiplication problem is 9.
5. (a) The correct answer is 37; this is another basic subtraction problem. A mistaken answer is likely due to an error in computation.
6. (c) The correct answer to this basic addition problem is 69. Again, a mistaken answer is likely due to an error in computation.
7. (b) The correct answer is 41.
8. (a) The correct answer is 45.
9. (d) The correct answer is 21.
10. (c) The correct answer is 36,597.
11. (b) In this case (but not in all cases, as you'll see in some problems below), perform the operations in the order presented, from left to right: 72 + 98 = 170; then, 170 − 17 = 153.
12. (d) Again, do this problem in the order presented; first subtract, then add. 353 is the correct answer.
13. (c) First add, then subtract. In multi-step problems, be careful not to rush just because the operations are simple. The correct answer is 560.
14. (b) First add, then subtract. The correct answer is 6,680. When doing problems involving several digits, it's easy to make mistakes. Be sure to read carefully, no matter how simple the problem seems.
15. (a) First subtract, then add. The correct answer is 5,507.
16. (d) Subtract twice: 1,556.

1000 MATH PROBLEMS >>> Answers

SET 2

17. (d) Perform the operations within the parentheses first: $25 + 17 = 42$; $64 - 49 = 15$. Then go on to multiply those answers as indicated by the two side-by-side parentheses: $42 \times 15 = 630$.

18. (b) The correct answer is 7.

19. (a) The correct answer to this multiplication problem is 14,600. An incorrect answer is likely an error in computation, particularly in not carrying digits to the next place.

20. (c) The correct answer to this multiplication problem is 30,400. Again, an incorrect answer may result from the common error of not carrying digits to the next place.

21. (a) The correct answer is 9,092.

22. (a) The correct answer (rounded to the nearest whole number) is 5,870.

23. (a) The correct choice is 5,001. If you got a different answer, you may have misread the answer choices because they appear so similar. It is important to read carefully even when you are in a hurry.

24. (c) The correct answer is 256,595. When multiplying three-digit numbers, you have to be especially careful to align numbers.

25. (a) 15 is the correct answer. An incorrect answer may represent an error in place value.

26. (c) The correct answer is 12,407. If you got answer a, you disregarded the zero in 62,035.

27. (c) The correct answer is 100.12, rounded to 100.

28. (a) The correct answer is 205,335

29. (d) Again, perform the operations in parentheses first: First, do the math that is in the second parentheses, because it is multiplication: $3 \times 54 = 162$. Now do the math in the first parentheses: $84 - 5 = 79$, then multiply that result by 12: $12(79) = 948$. Now do the final operation: $948 - 162 = 786$.

30. (c) Perform the operation in parentheses first: $2 + 4 = 6$. (Even though this part is addition, you do it first because it is in parentheses.) Now multiply: $6 \times 8 = 48$. If you picked choice b, you probably multiplied the numbers in parentheses, rather than adding them.

31. (c) Perform the operation in parentheses first: $14 \times 7 = 98$, and then add 12 to get the answer, which is 110.

32. (a) First multiply, then add. The correct answer is 1,467.

1000 MATH PROBLEMS >>> Answers

SET 3

33. (c) First multiply, then subtract : The correct answer is 65,011.

34. (d) Divide first, then add : The correct answer is 101.

35. (a) In spite of the sequence presented, in this problem, you must divide first : 204 ÷ 2 = 102. Now subtract the solution from 604: 604 − 102 = 502. The correct answer is 502. When an expression involves more than one operation, do the operations in the following order: (1) Operations in Parentheses; (2) Exponents; (3) Multiplication; (4) Division; (5) Addition; (6) Subtraction. A good device for remembering the order is this saying: Please excuse **my dear Aunt Sally.**

36. (d) Do the operation in parentheses first: 202 ÷ 2 = 101; then subtract the answer from 604: 604 − 101 = 503.

37. (d) Multiply twice. The correct answer is 6,660.

38. (a) It is a good idea to develop the habit of always performing operations in parentheses first, even though in this case it doesn't matter because all the operations are multiplication. Multiply 9 × 4 = 36, then multiply again: 36 × 12 = 432.

39. (d) Saying the numbers out loud helps: Twenty thousand plus seven hundred plus six. Choice **a** reads 276; choice **b** reads 2,706; choice **c** reads 20,076.

40. (c) Choice **a** is not divisible by 6; choice **b** is not divisible by 6 or 7; and choice **d** is not divisible by 7. 84 is divisible by both numbers: 6 × 7 × 2 = 84.

41. (a) The expression $5n$ means 5 times n. The addition sign before the 7 indicates the phrase *more than.*

42. (a) 46 goes into 184 four times. The other choices cannot be divided evenly into 184.

43. (c) Subtract the pounds first, then the ounces.

44. (d) You must "borrow" 60 minutes from the 3 hours in order to be able to subtract.

45. (c) Add the four numbers together to get 260, then divide by 4 to get 65.

46. (a) If you got a different answer, you probably made an error in your division.

47. (b) Add the feet first, then the inches: 2 feet + 4 feet = 6 feet. 4 inches + 8 inches = 12 inches. Convert 12 inches into 1 foot to get the correct answer 6 feet + 1 foot = 7 feet.

48. (c) Add the hours first, then the minutes: 1 hour + 3 hours = 4 hours. 20 minutes + 30 minutes = 50 minutes. Combine: 4 hours 50 minutes.

SET 4

49. (a) 157 is rounded to 200; 817 is rounded to 800. 200 × 800 = 160,000.

50. (c) When *adding* negative numbers, follow this rule: If both numbers have **different** signs, subtract the smaller number from the larger. The answer has

1000 MATH PROBLEMS >>> Answers

the sign of the larger number. Therefore, the above equation becomes: 12 − (**not** +) 4 = 8. Since the larger number, 12, has a negative sign, the answer has a negative sign, so the answer is −8.

51. (d) When *multiplying* negative numbers, begin by simply multiplying the two numbers together. If both numbers have the **same sign** (whether plus or minus), the answer is positive; otherwise, the answer is negative. So: 4 × 6 = 24. Both numbers have the same sign (in this case negative), so the answer is positive: 24.

52. (d) When *adding* negative numbers, if both numbers have the **same sign**, begin by simply adding. If the sign of both numbers is negative, the answer is negative.

53. (b) You are adding numbers that have **different signs**; therefore, subtract the smaller from the larger: 10 − 6 = 4. Now give the answer the same sign as the larger number, which in this case is a negative number, so the answer is −4.

54. (a) When adding two numbers that are the **same**, if the two numbers have **opposite** signs, the answer is zero.

55. (c) The meaning of 4^3 is 4 *to the power of 3*, or 4 times itself 3 times.

56. (d) The superscript 4 means *to the power of 4*; in other words, take 3 times itself four times: 3 × 3 × 3 × 3 = 81.

57. (d) The exponent here is 3, which is the power to which the number is raised—that is, 6^3 = 6 times itself 3 times, or: 6 × 6 × 6 = 216.

58. (d) 17^2 means 17 squared and is equivalent to 17 × 17, which equals 289.

59. (b) To solve this division problem, *subtract* the exponents only: 5 − 2 = 3, so the answer is 10^3.

60. (b) Minus 6^2 is equal to −6 multiplied by itself, or −6 × (−6). When multiplying two negative numbers, the answer is positive: 36.

61. (c) Minus 3 to the third power is −3 multiplied by itself 3 times. (−3) × (−3) × (−3) = −27.

62. (a) Minus 12^2 is 12 times itself. −12 × (−12) = 144. Because the signs of the numbers are the same, the answer is positive.

63. (c) To find the square root of a number, ask yourself, "What number times itself equals the given number?" Four times itself, or 4^2, is 16; therefore, the square root of 16 is 4.

64. (a) Square roots can be multiplied and divided, but they cannot be added or subtracted.

SET 5

65. (c) The square root of 64 is 8.

66. (a) Divide the total amount of silver (6,000 kg) by the amount in each bag (5 kg), to get the number of bags (1,200).

1000 MATH PROBLEMS >>> Answers

67. (d) Choice **d** is in the correct order. Choice **a** omits information (Spike), and choices **b** and **c** present inaccurate information.

68. (b) This is a basic addition problem. Begin by adding: 15 metres + 30 centimetres = 35 metres 85 centimetres. Since there are 60 centimetres in a metre, 85 centimetres becomes 1 metre 25 centimetres.

69. (b) Change the hours to minutes: 1 hour 40 minutes = 100 minutes; 1 hour 50 minutes = 110 minutes. Now add: 100 minutes + 110 minutes = 210 minutes. Now change back to hours and minutes: 210 minutes ÷ 60 = 3.5 hours.

70. (b) This is a problem in multiplication: 15 (number of people who brought cats) × 3 (number of cats) = 45 (cats).

71. (b) To answer this question, set up this equation: Rs. 47 − Rs. 5 − Rs. 15 = ?, and do the operations in the order presented. The correct answer is Rs. 27.

72. (b) Add all four weights for a total of 703 reels.

73. (c) When the values are added together, the amount stolen was Rs. 7050. (The two rings valued at Rs. 1500 have a total value of Rs. 3000 but remember that there is another ring valued at only Rs. 700.

74. (c) You must subtract the reading at the beginning of the week from the reading at the end of the week: 21,053 − 20,907 is 146.

75. (c) The values added together total Rs. 618. If you chose option a, you forgot that the value of the handbag (Rs. 150) must also be included in the total.

76. (b) This is a basic addition problem: 108 kgs + 27 kgs = 135 kgs.

77. (c) The total value is Rs. 5,525. It is important to remember to include all three hot cases (Rs. 375 total), both flasks (Rs. 2,600 total), and both shelfs (Rs. 1,900 total) in the total value.

78. (b) Add the corrected value of the sweater (Rs. 245) to the value of the two, not three, bracelets (Rs. 730), plus the other two items (Rs. 78 and Rs. 130) to get the answer, Rs. 1183.

79. (a) This is a two-step subtraction problem. First you must find out how many miles the truck has traveled since its last maintenance. To do this subtract: 22,003 − 12,398 = 9,605. Now subtract 9,605 from 13,000 to find out how many more miles the truck can travel before it must have another maintenance: 13,000 − 9,605 = 3,395.

80. (d) This is a problem of multiplication. The easiest way to solve this problem is to temporarily take away the five zeros, then multiply: 365 × 12 = 4,380. Now add back the five zeros for a total of 438,000,000. (If you chose option a, you mistakenly divided rather than multiplied.)

SET 6

81. (c) First ask how many inches are in one foot; the answer is 12 inches. Now multiply: 12 × 4 = 48 inches.

1000 MATH PROBLEMS >>> Answers

82. (c) This is basic division problem: 46 ÷ 2 = 23. Rajeeve is 23 years old.

83. (a) This is a two-step division problem: 2,052 miles ÷ 6 days = 342 miles per day. 342 miles per day ÷ 2 stops = 171 miles between stops.

84. (b) Shaloo's three best (that is, shortest) times are 54, 54, and 57, which add up to 165. Now divide to find the average of these times: 165 ÷ 3 = 55. If you got the wrong answer, you may have added all of Shaloo's times, rather than just her best three. Even when the problem seems simple and you're in a hurry, be sure to read carefully.

85. (d) To find the average, divide the total number of miles, 3,450, by 6 days: 3,450 miles ÷ 6 days = 575 miles per day.

86. (c) First find the total number of patients by adding: 8 + 5 + 9 + 10 + 10 + 14 + 7 = 63. Then find the average by dividing the number of patients by the number of nursing assistants: 63 ÷ 7 = 9.

87. (d) Take the total number of miles and find the average by dividing: 448 miles ÷ 16 gallons = 28 miles per gallon.

88. (d) 827, 036 bytes free + 542,159 bytes freed when the document was deleted = 1,369,195 bytes. 1,369,195 bytes — 489,986 bytes put into the new file = 879,209 bytes left free.

89. (c) The solution to the problem entails several operations: First, multiply Rs. 80 per month by 7 months = Rs. 560. Next, multiply Rs. 20 per month by the remaining 5 months = Rs. 100. Now find the average for the entire year. Add the two amounts: Rs. 560 + Rs. 100 = Rs. 660. Now divide: Rs. 660 ÷ 12 months in a year = Rs. 55.

90. (c) This is a two-step problem. First, add the three numbers: 22 + 25 + 19 = 66. Now divide the sum by 3 to find the average: 66 ÷ 3 = 22.

91. (c) First, convert feet to inches: 3 feet = 3 × 12 inches = 36 inches. Now add 4 inches: 36 inches + 4 inches = 40 inches. Then do the final operation: 40 inches ÷ 5 = 8 inches.

92. (c) Multiply 16 × 5 to find out how many gallons all five sprinklers will release in one minute. Then multiply the result (80 gallons per minute) by the number of minutes (10) to get the entire amount released: 80 × 10 = 800 gallons.

93. (b) The median value is the middle value when the numbers are sorted in descending order. The answer here is 10 inches.

94. (d) It will take one worker about twice as long to complete the task, so you must multiply the original hours and minutes times 2: 2 hours 40 minutes × 2 = 4 hours 80 minutes, which is equal to 5 hours 20 minutes.

95. (b) Two candy bars require 2 quarters; one package of peanuts requires 3 quarters; one can of cola requires 2 quarters—for a total of 7 quarters.

96. (b) He spent 2.2 hour 20 minutes before 12 noon and 4 hours 15 minutes at after noon. Total 6 hours 35 minutes. If 30 minutes is spent in coming and going remaining time is 6 hours 5 minutes.

1000 MATH PROBLEMS >>> Answers

SET 7

97. (a) To arrive at the answer quickly, begin by rounding off the numbers, and you will see that choice **a** is less than 300 kgs, whereas choices b, c, and d are all over 300 kgs.

98. (b) Take the number of classroom hours and divide by the number of courses: $48 \div 3 = 16$ hours per course. Now multiply the number of hours taught for one course by the pay per hour: $16 \times$ Rs. $35 =$ Rs. 560.

99. (b) Anita worked 15 hours per week for 8 weeks: $15 \times 8 = 120$. In addition, she worked 15 hours for Manju for one week, so: $120 + 15 = 135$.

100. (a) Divide the number of frees by the number of garden $11376 \div 48 = 237$.

101. (c) It is best to make the problem as simple and clear as possible, so B = 46 is the best choice. Choice **a** is not simplified as much as it could be; choice **b**, although technically correct, is less clear and more apt to cause confusion if stated in this order; choice **d** is inaccurate.

102. (b) The average is the sum divided by the number of observations: $(11 + 4 + 0 + 5 + 4 + 6) \div 6 = 5$.

103. (a) The labour fee (Rs. 25) plus the deposit (Rs. 65) plus the basic service (Rs. 15) equals Rs. 105. The difference between the total bill, Rs. 112.50, and Rs. 105 is Rs. 7.50, the cost of the news channels.

104. (c) 3 tons = 3,000 kg. 3,000 kg × 1,000 grams per kg = 3,000,000 g that can be carried by the truck. The total weight of each daily ration is: 200 g + 275 g + 275 g = 750 g per soldier per day. 3,000,000 ÷ 750 = 2,000. 2,000 ÷ 10 days = 200 soldiers supplied.

105. (c) Between 8:14 and 9:00, 46 minutes elapse, and between 9:00 and 9:12, 12 minutes elapse, so this is a simple addition problem: $46 + 12 = 58$.

106. (d) Begin by multiplying to find out how many forms one clerk can process in a day: 26 forms × 8 hours = 208 forms per day per clerk. Now find the number of clerks needed by dividing: $5,600 \div 208 = 26.9$. Since you can't hire 0.9 of a clerk, you have to hire 27 clerks for the day.

107. (a) The companies combined rate of travel is: 35 miles per hour + 15 miles per hour = 50 miles per hour. 2,100 miles ÷ 50 miles per hour = 42 hours.

108. (a) This answer is in the correct order and is "translated" correctly: Ravi had (=) 3 apples and ate (−) 1.

109. (c) This is a multiplication problem: 84,720 troops × 4 metres of cloth equals 338,880 metres of cloth required. (If you chose option a, you mistakenly divided.)

110. (c) Subtraction and addition will solve this problem. From 10:42 to 12:42 two hours have elapsed. From 12:42 to 1:00, anther 18 minutes have elapsed (60 − 42 = 18). Then from 1:00 to 1:19, another 19 minutes have elapsed. Now add: 2 hours + 18 minutes + 19 minutes = 2 hours 37 minutes.

1000 MATH PROBLEMS >>> Answers

111. (c) To solve this problem, multiply the amount saved per fire Rs. 5,700, by the average number of fires: 5,700 × 14 = 79,800.

112. (c) According to the clinic's scale, a foot in height makes a difference of 60 kgs, or 5 kgs per inch of height over 5 feet. A person who is 5 feet 5 inches tall should therefore be no more than (5) (5 kgs), or 25 kgs, heavier than the person who is 5 feet tall. So, add: 25 kgs + 80 kgs = 105 kgs.

SET 8

113. (b) To find the answer, begin by adding the cost of the two sale puppies Rs. 15 + Rs. 15 = Rs. 30. Now subtract this amount from the total cost Rs 70 − Rs 30 = Rs. 40 paid for the third puppy.

114. (c) January is approximately Rs 38,000; February is approximately 41,000, and April is approximately 26,000. These added together give a total of 105,000.

115. (d) Because Laxmi only pays with a cheque if an item costs more than Rs. 30, the item Laxmi purchased with a cheque in this problem must have cost more than Rs. 30. If an item costs more than Rs. 30, then it must cost more than Rs. 25 (choice **d**), as well.

116. (d) This series actually has two alternating sets of numbers. The first number is doubled, giving the third number. The second number has 4 subtracted from it, giving the fourth number. Therefore, the blank space will be 12 doubled, or 24.

117. (d) Figure the amounts by setting up the following equations: First, S = Rs. 3 + Rs. 23 = Rs. 26. Now B = (Rs. 1 × 5) + (Rs. 2 × 2) or Rs. 5 + Rs. 4 = Rs. 9. MR = Rs. 1 × 2 = Rs. 2. D = Rs. 4 × 1 = Rs. 4. Now, add: Rs. 9 + Rs. 2 + Rs. 4 = Rs. 15. Now subtract: Rs. 26 − Rs. 15 = Rs. 11.

118. (d) The values totaled are equal to Rs. 8,005. Be sure to read the complete question—that is, don't forget to add the value of the watch and the sheet music.

119. (c) Do the operations in chronological order: Begin with the first 7 people who plan to go on the trip (don't forget the pilot): Now subtract Bharat and Soni: 7 − 2 = 5. Now add Rama: 5 + 1 = 6. Now take away Ashi: 6 − 1 = 5.

120. (a) It is easiest to use trial and error to arrive at the solution to this problem. Begin with choice **a**: After the first hour, the number would be 20, after the second hour 40, after the third hour 80, after the fourth hour 160, and after the fifth hour 320. Fortunately, in this case, you need go no further. The other answer choices do not have the same outcome.

121. (a) The unreduced ratio is 8,000:5,000,000; reduced, the ratio is 8:5000. Now divide: 5000 ÷ 8 = 625, for a ratio of 1:625.

122. (c) This is a three-step problem involving multiplication, subtraction, and addition. First find out how many fewer minutes Girish jogged this week than usual:

1000 MATH PROBLEMS >>> Answers

5 hours × 60 minutes = 300 minutes − 40 minutes missed = 260 minutes jogged. Now add back the number of minutes Girish was able to make up: 260 minutes + 20 + 13 minutes = 293 minutes. Now subtract again: 300 minutes − 293 = 7 minutes jogging time lost.

123. (b). To find the average, divide the total number of people by the number of days Toni drives: 300 ÷ 15 = 20.

124. (c) This is a multiplication problem: Each front foot along the lake costs Rs. 250, so multiply; Rs. 250 × 300 feet = Rs. 75,000.

125. (d) First, write the problem in columns:

6 feet	5 inches
−5 feet	11 inches

Now subtract, beginning with the right-most column. Since you cannot subtract 11 inches from 5 inches, you must borrow one foot from the 6 in the top left column, then convert it to inches and add: 1 foot = 12 inches; 12 inches + 5 inches = 17 inches. The problem then becomes:

5 feet	17 inches
−5 feet	11 inches
	6 inches

So the answer is choice **d**, 6 inches.

126. (c) This is a problem of addition. You may simplify the terms: M = F + 10, then substitute: M = 16 + 10, or 26.

127. (c) 76 ÷ 19 = 4. The other division operations will not end in whole numbers.

128. (b) The mode is the number that appears most frequently in a series—in this case, it is 8.

SET 9

129. (a) Subtract the months first, then the years. Remember that it is best to write the problem in columns and subtract the right-most column (months) first, then the left-most column (years): 8 months − 7 months = 1 month; 2 years − 1 year = 1 year. So, Beena is 1 year 1 month older than Minu.

130. (b) First, simplify the problem: K = 50 × 17 = 850 so Kara made 850 paise. R = 75 × 14 = 1050, so Rani made 1050 paise, so Rani made the most money. R − K = 1050 − 850 = 200. Therefore, Rani made 200 paise more than Kara.

131. (a) In this problem you must find an average. So divide the total number of earnings and divide by the number of months 51858 ÷ 12 = Rs. 4321.50

132. (b) First, divide to determine the number of 20 minute segments there are in an hour: 60 ÷ 20 = 3. Now multiply that number by the number of times Rita can circle the garden 3 × 5 = 15.

1000 MATH PROBLEMS >>> Answers

133. (a) The symbol > means "greater than," and the symbol < means "less than". The only sentence that is correct is choice **a**: 4 feet is greater than 3 feet. The other choices are untrue.

134. (c) A prime number is one that can be divided evenly by itself and 1, but not by any other number. The other choices are divisible by other numbers, besides 1 and themselves.

135. (d) First, change the names to letters; remember that the letters then represent, not the people, but their *ages*. S (Soni's age) equals R (Rama's age) plus 10 (years).

136. (c) The *average* (the sum of a group of numbers divided by the number of numbers) is also called the *mean*.

137. (c) First, simplify the problem: L = Rs. 5 + Rs. 4 + Rs. 6 = Rs. 15; R = Rs. 14. Lalith spent the most by Rs. 1. Don't forget that only the popcorn was shared; the other items must be multiplied by 2.

138. (d) First, convert tons to kgs. 1 ton = 1,000 kg. 36 tons (per year) = 36,000 kg (per year). 1 year = 12 months, so the average number of kgs of mosquitoes the colony of bats can consume in a month is: 36,000 ÷ 12, or 3,000 kgs.

139. (a) 10^{18} means 10 to the 18th power, or 10 times itself 18 times.

140. (c) This is a three-step problem involving multiplication and division. First, change yards to feet: 240 yards × 3 feet in a yard = 720 feet. Now find the number of square feet in the parcel: 121 feet × 740 feet = 87,120 square feet. Now find the acres 87,120 ÷ 43,560 square feet in an acre = 2 acres.

141. (a) This is a multiplication problem. A quarter section contains 160 acres, so you must multiply: 160 × Rs. 1,850 = Rs. 296,000.

142. (c) This is a subtraction problem. First, simplify the problem by dropping the word "million." The problem then becomes P = 3,666, M = 36. So P − M = 3,666 − 36 = 3,630. Now add the word "million" back and the answer becomes 3,630 million.

143. (b) 860 feet × 560 feet ÷ 43,560 square feet per acre = 11.05 acres.

144. (b) 30 men × 42 square feet = 1,260 square feet of space. 1,260 square feet ÷ 35 men = 36 square feet. 42 − 36 = 6, so each man will have 6 less square feet of space.

SET 10

145. (b) The difference between 105 and 99 is 6 degrees. Application of the ice pack plus a "resting" period of 5 minutes before reapplication means that the temperature is lowered by half a degree every six minutes, or 1 degree every 12 minutes. 6 degrees × 12 minutes per degree = 72 minutes, or 1 hour and 12 minutes.

146. (b) The median is merely the number in the middle of the series, which in this case is 12.

147. (b) Since there are two middle numbers in this set, 14 and 16, the median is the average of the two, or 15.

1000 MATH PROBLEMS >>> Answers

148. (a) You cannot subtract square roots; however, it is possible to do this problem, because you don't have to deal with the "squares" at all, just subtract. That is, the way the problem is set up, you could substitute any noun in place of "$\sqrt{3}$." As used in the problem, it is simply a unit with no mathematical meaning for this particular operation. (If you had 4 "*blogs*" and took away 2 "*blogs*", what you would have left is 2 "*blogs*.")

149. (b) $\sqrt{12}$ is the same as $\sqrt{4 \times 3}$, which is the same as $\sqrt{4} \times \sqrt{3}$. The square root of 4 is 2. So $3 \times \sqrt{12}$ is the same as $3 \times 2 \times \sqrt{3}$.

150. (c) 79 goes into 237, 3 times.

151. (c) Take the words in order and substitute the letters and numbers: *Rinku has* (=) *3 times* (×) *the number of tennis trophies Avinash has* "translates" to R = 3A.

152. (c) The correct answer here is 10,447. It helps, if you are in a place where you can do so, to read the answer aloud; that way, you'll likely catch any mistake. When writing numbers with more than 4 digits, begin at the right and separate the digits into groups of three with commas.

153. (c) 10,043,703 is the correct answer. The millions place is the third group of numbers from the right. If any group of digits *except the first* has less than 3 digits, you must add a zero at the beginning of that group.

154. (c) This is a several-step problem. First figure out how many *dozen* eggs John's chickens produce per day: 480 ÷ 12 = 40 dozen eggs per day. Now figure out how much money John makes on eggs per day: Rs. 2.00 × 40 = Rs. 80 per day. Finally, figure out how much money John makes per week: Rs. 80 × 7 = Rs. 560 per week. The most common mistake for this problem is to forget to do the last step. It is important to read each problem carefully, so you won't skip a step.

155. (c) Because of the relatively small, even numbers that end in zeros, this is among the simplest kinds of division problems. Do such problems in your head in order to save time for more complicated ones: 60 ÷ 10 = 6 kgs.

156. (b) This is a basic division problem, which you might want to do in your head, in order to save time for more complicated ones: 27 ÷ 3 = 9.

157. (d) The production for Lunar's is equal to the total minus the other productions: 1780 − 450 − 425 − 345 = 560.

158. (b) 54 divided by 6 is 9.

159. (a) The mean is equal to the sum of values divided by the number of values. Therefore, 8 raptors per day × 5 days = 40 raptors. The sum of the other six days is 34 raptors. 40 raptors − 34 raptors = 6 raptors.

160. (c) The speed of the train, obtained from the table, is 60 miles per hour. Therefore, the distance from Delhi would be equal to $60t$. However, as the train moves on, the distance decreases from Patna, so there must be a function of $-60t$ in the equation. At time $t = 0$, the distance is 2000 miles, so the function is $2000 - 60t$.

SECTION 2 : FRACTIONS

SET 11

161. (b) Two of the four sections are shaded, so $\frac{2}{4}$ of the figure is shaded. Reducing, the answer is $\frac{1}{2}$.

162. (b) The fraction is the same, no matter which parts of the figure are shaded. 3 parts out of 5, or $\frac{3}{5}$.

163. (a) In order to subtract fractions, you must first find the least common denominator, which in this case is 40. After conversion, the equation is: $\frac{35}{40} - \frac{24}{40} = \frac{11}{40}$.

164. (c) The mixed numbers must first be converted to fractions, and you must use the least common denominator, which in this case is 8. The equation then becomes: $\frac{18}{8} + \frac{37}{8} + \frac{4}{8} = \frac{59}{8}$. Now reduce: $\frac{59}{8} = 7\frac{3}{8}$.

165. (a) To arrive at a solution, first convert the mixed number to fractions: $\frac{37}{10} - \frac{19}{8}$. Next, find the least common denominator, which is 40. So, $\frac{148}{40} - \frac{95}{40} = \frac{53}{40}$. Finally, reduce $\frac{53}{40} = 1\frac{13}{40}$.

166. (d) First, find the least common denominator—that is, convert all three fractions to twelfths, then add: $\frac{2}{12} + \frac{7}{12} + \frac{8}{12} = \frac{17}{12}$. Now reduce: $1\frac{5}{12}$.

167. (d) You must convert both fractions to thirtieths before adding: $4\frac{10}{30} + 3\frac{9}{30} = 7\frac{19}{30}$.

168. (a) First, convert the mixed numbers to fractions. Since any number multiplied by 1 retains its identity, we have: $3\frac{9}{16} = (3)(\frac{16}{16}) + \frac{9}{16} = \frac{48}{16} + \frac{9}{16} = \frac{57}{16}$ and $1\frac{7}{8} = (1)(\frac{8}{8}) + \frac{7}{8} = \frac{8}{8} + \frac{7}{8} = \frac{15}{8}$. Next, find the least common denominator of the two numbers, in this case 16, and convert: $(\frac{15}{8})(\frac{2}{2}) = \frac{30}{16}$. Finally, perform the indicated operation: $\frac{57}{16} - \frac{30}{16} = \frac{27}{16}$ which is equivalent to $(1)(\frac{16}{16}) + \frac{11}{16} = 1\frac{11}{16}$.

1000 MATH PROBLEMS >>> Answers

169. **(c)** First, convert the mixed numbers to fractions. We have: $4\frac{2}{5} = \frac{20}{5} + \frac{2}{5} = \frac{22}{5}$; and $3\frac{1}{2} = \frac{6}{2} + \frac{1}{2} = \frac{7}{2}$; and $\frac{3}{8}$. Next, find the least common denominator of the three numbers, in this case 40, and convert: $\frac{22}{5} = (\frac{22}{5})(\frac{8}{8}) = \frac{176}{40}$; $\frac{7}{2} = (\frac{7}{2})(\frac{20}{20}) = \frac{140}{40}$; $\frac{3}{8} = (\frac{3}{8})(\frac{5}{5}) = \frac{15}{40}$. Finally, perform the indicated operations: $\frac{176}{40} + \frac{140}{40} + \frac{15}{40} = \frac{331}{40}$, or $= 8\frac{11}{40}$.

170. **(b)** Before subtracting, you must convert both fractions to twenty-fourths: $\frac{10}{24} - \frac{9}{24} = \frac{1}{24}$.

171. **(d)** First, find the common denominator, which is 24. $\frac{3}{8} = \frac{9}{24}$; $\frac{5}{6} = \frac{20}{24}$. Then convert the mixed numbers to fractions and subtract: $\frac{1353}{24} - \frac{260}{24} = \frac{1093}{24}$. Now change back to a mixed number: $45\frac{13}{24}$.

172. **(a)** Remember that two negatives make a positive. Find the common denominator ($\frac{3}{10}$ and $\frac{1}{5}$ are equal to $\frac{3}{10}$ and $\frac{2}{10}$). Next, invert: $\frac{3}{10} \div \frac{2}{10}$ becomes $\frac{3}{10} \times \frac{10}{2} = \frac{30}{20}$ or $1\frac{1}{2}$.

173. **(c)** The correct answer is 12. One of the most common errors is found in answer **(d)**, where the numbers were multiplied rather than divided.

174. **(b)** The correct answer is $1\frac{1}{6}$.

175. **(a)** Again, in order to subtract the fractions, you must first find the least common denominator, which in this case is 40. The equation is then $\frac{35}{40} - \frac{24}{40} = \frac{11}{40}$.

176. **(b)** Convert to the lowest common denominator, which is 10, then add. The correct answer is $8\frac{9}{10}$. (Incorrect answers may result from adding both the numerator and the denominator and from failing to convert fifths to tenths properly.)

SET 12

177. **(b)** To work the problem you must first convert $\frac{1}{2}$ to $\frac{3}{6}$, then add. The correct answer is $88\frac{1}{3}$.

178. **(c)** This is a basic problem of subtraction. Common errors include **(d)**, misreading the subtraction sign and adding the fractions instead; or **(a)**, subtracting the numerators but adding the denominators.

1000 MATH PROBLEMS >>> Answers

179. (c) First, find the least common denominator of the fractions, which is 9, then add the fractions: $\frac{6}{9} + \frac{3}{9} = \frac{9}{9}$ or 1. Now add the whole numbers: $43 + 36 = 79$. Now add the results of the two equations: $1 + 79 = 80$.

180. (d) The lowest common denominator is 24. $\frac{15}{24} - \frac{8}{24} = \frac{7}{24}$.

181. (d) The correct answer is $\frac{58}{35}$. (Unlike the other choices, this could be changed to mixed number if desired.)

182. (b) $\frac{3}{7}$ and $\frac{7}{3}$ are reciprocals. The product of reciprocals is always 1.

183. (c) Because two negatives equal a positive, answers (a) and (d) can be easily ruled out. Answer (b) is the result of multiplying rather than dividing.

184. (b) One common error, dividing the fractions instead of multiplying, is found in answer (d).

185. (a) The first step is to convert the number 3 to a fraction, making it $\frac{3}{1} \cdot \frac{5}{8} \div \frac{3}{1}$ is accomplished by inverting the second fraction, making it $\frac{1}{3}$, and multiplying: $\frac{5}{8} \times \frac{1}{3} = \frac{5}{24}$.

186. (b) First, change the mixed numbers to improper fractions. (That is, for each fraction multiply the whole number by the denominator of the fraction, then add the numerator: $2 \times 4 + 1 = 9$, so the fraction becomes $\frac{9}{4}$. $2 \times 7 + 4 = 18$, so the fraction, becomes $\frac{18}{7}$. The equation then becomes $\frac{9}{4} \div \frac{18}{7}$. Now invert the second fraction and multiply the numerators and the denominators: $\frac{9}{4} \times \frac{7}{18} = \frac{63}{72}$. Reducing, the answer becomes $\frac{7}{8}$. (If you performed a multiplication operation instead of a division one, you got wrong answer (d).)

187. (d) You can cancel before multiplying, or simply multiply all the numerators and all the denominators to get $\frac{12}{100}$. Then reduce to get $\frac{3}{25}$.

188. (b) Convert the fractions to mixed numbers: $1\frac{1}{2} = \frac{3}{2}$, and $1\frac{5}{13} = \frac{18}{13}$. Now invert the second fraction and multiply: $\frac{3}{2} \times \frac{13}{18} = \frac{39}{36}$ or $1\frac{3}{36}$. Now reduce: $1\frac{1}{12}$. (A common error is shown in (c), multiplying fractions instead of dividing.)

189. (d) Properly converting the mixed numbers into improper fractions is the first step in finding the answer. Thus $\frac{7}{3} \times \frac{15}{14} \times \frac{9}{5} = \frac{945}{210} = 4\frac{1}{2}$.

1000 MATH PROBLEMS >>> *Answers*

190. **(a)** To multiply fractions, you must multiply the numerators to reach the numerator of the answer (2 × 3 = 6) and multiply the denominators to reach the denominator of the answer (5 × 7 = 35). So the correct answer is $\frac{6}{35}$.

191. **(c)** First, change $2\frac{1}{4}$ to an improper fraction: $2\frac{1}{4} = \frac{9}{4}$. Next, in order to divide by $\frac{2}{3}$, invert that fraction to $\frac{3}{2}$ and multiply: $\frac{9}{4} \times \frac{3}{2} = \frac{27}{8} = \frac{(24+3)}{8} = \frac{24}{8} + \frac{3}{8} = 3\frac{3}{8}$.

192. **(a)** For the answer, divide $\frac{2}{3}$ by $\frac{5}{12}$, which is the same as $\frac{2}{3} \times \frac{12}{5} = \frac{24}{15} = 1\frac{3}{5}$.

SET 13

193. **(a)** To multiply a whole number by a fraction, first change the whole number to a fraction (that is, place it over 1). Then multiply the fractions: $\frac{4}{1} \times \frac{1}{3} = \frac{4}{3}$ or $1\frac{1}{3}$.

194. **(d)** The correct answer is 4.

195. **(b)** If you multiplied instead of dividing, you got answer **(d)**.

196. **(b)** This is a basic division of fractions problem. First, covert the mixed number to a fraction: $\frac{21}{8} \div \frac{1}{3} \cdot \frac{21}{8} \times \frac{3}{1} = \frac{63}{8}$, or $7\frac{7}{8}$.

197. **(b)** A common error is shown in **(c)**, multiplying fractions instead of dividing.

198. **(d)** Properly converting the mixed numbers into improper fractions is the first step in finding the answer.

199. **(c)** The correct answer is $30\frac{2}{5}$.

200. **(b)** The correct answer is $\frac{12}{15}$, or $\frac{4}{5}$.

201. **(d)** The first step in solving this problem is to subtract to get $-\frac{6}{3}$. This reduces to −2.

202. **(b)** The correct answer is $\frac{7}{32}$.

203. **(a)** The correct answer is $18\frac{2}{3}$.

204. **(a)** The least common denominator for the two fractions is 18. Ignore the whole number 6 for a moment, and simply subtract the fractions: $\frac{4}{18} - \frac{3}{18} = \frac{1}{18}$. Now put back the 6 for the correct answer: $6\frac{1}{18}$. (If you got answer **(c)**, you subtracted both the numerators and the denominators.)

1000 MATH PROBLEMS >>> Answers

205. (c) The correct answer is $13\frac{29}{35}$.

206. (a) The correct answer is $17\frac{8}{9}$.

207. (d) Find the answer by changing the fractions to decimals: $\frac{1}{3} = 0.333$; $\frac{1}{4} = 0.25$; $\frac{2}{7} = 0.286$. 0.286, or $\frac{2}{7}$, is between the other two.

208. (b) Divide the bottom number into the top number $17 \div 3 = 5\frac{2}{3}$.

SET 14

209. (a) Divide the numerator of the fraction, or top number, by the denominator of the fraction, or bottom number: So $\frac{13}{25}$ becomes $13 \div 25$, or 0.52, or $\frac{52}{100}$.

210. (a) To solve this problem, you must first convert all the fractions to the lowest common denominator, which is 24. $\frac{7}{8} = \frac{21}{24}$; $\frac{3}{4} = \frac{18}{24}$; $\frac{2}{3} = \frac{16}{24}$; $\frac{5}{6} = \frac{20}{24}$.

211. (b) Fractions must be converted to the lowest common denominator, which is 60. $\frac{6}{10} = \frac{36}{60}$; $\frac{11}{20} = \frac{33}{60}$; $\frac{8}{15} = \frac{32}{60}$, which is the smallest fraction.

212. (d) To solve the problem, one must first find the common denominator, in this instance 60. Then the fractions must be converted: $\frac{17}{20} = \frac{51}{60}$ (for choice **(a)**); $\frac{3}{4} = \frac{45}{60}$ (for choice **(b)**); $\frac{5}{6} = \frac{50}{60}$ (for choice **(c)**); and $\frac{7}{10} = \frac{42}{60}$ or the correct choice, **(d)**.

213. (b) Convert the mixed number $3\frac{7}{8}$ to the improper fraction $\frac{31}{8}$ and then invert.

214. (a) Change the mixed number to an improper fraction: $3\frac{3}{4} = \frac{15}{4}$. Now invert: $\frac{4}{15}$.

215. (d) Divide the top number by the bottom number. $160 \div 40 = 4$.

216. (b) Multiply the numerator by the denominator: $15 \div 2 = 7\frac{1}{2}$.

217. (a) Multiply the whole number by the fraction's denominator. $5 \times 2 = 10$. Add the fraction's numerator to the answer: $1 + 10 = 11$. Now place that answer over the fraction's denominator: $\frac{11}{2}$.

218. (a) This is a simple subtraction of fractions problem. The first step is to find the common denominator, which is 15.

219. (b) In an improper fraction the top number is greater than the bottom number.

220. (a) Manish has finished $\frac{35}{45}$ of his test, which reduces to $\frac{7}{9}$, so he has $\frac{2}{9}$ of the test to go.

1000 MATH PROBLEMS >>> Answers

221. (a) First find the least common denominator of the two fractions, which is 6. Then add the fractions of the sandwich Jona got rid of: $\frac{3}{6}$ (which she gave to Mona) + $\frac{2}{6}$ (which she ate) = $\frac{5}{6}$. Now subtract the fraction from one whole sandwich (1 = $\frac{6}{6}$): $\frac{6}{6} - \frac{5}{6} = \frac{1}{6}$.

222. (a) The total width of the three windows is 105 inches. $105 \times 2\frac{1}{2} = \frac{105}{1} \times \frac{5}{2} = \frac{525}{2} = 262\frac{1}{2}$.

223. (b) There is $\frac{1}{3}$ of the sweet left after the first day. $\frac{1}{2}$ of $\frac{1}{3} = \frac{1}{2} \times \frac{1}{3} = \frac{1}{6}$.

224. (c) To subtract fractions, first convert to a common denominator, in this case, $\frac{25}{40} - \frac{24}{40} = \frac{1}{40}$.

SET 15

225. (c) To find out how many dozen cookies Ram can make, divide $5\frac{1}{2}$ by $\frac{2}{3}$. First, convert $5\frac{1}{2}$ to $\frac{11}{2}$, then multiply by $\frac{3}{2}$, which is the same as dividing by $\frac{2}{3}$. $\frac{11}{2} \times \frac{3}{2} = \frac{33}{4}$, or $8\frac{1}{4}$ dozen.

226. (b) The common denominator of the fractions is 280. The sum of the fractions is $\frac{503}{280}$, or $1\frac{223}{280}$. This unwidely fraction cannot be reduced further.

227. (a) To multiply mixed numbers, convert to improper fractions, or $\frac{5}{4} \times \frac{11}{8} = \frac{55}{32}$, or $1\frac{23}{32}$ square inches.

228. (d) Convert both numbers ($4\frac{1}{2}$ and 7) to improper fractions, to get: $\frac{9}{2} \div \frac{7}{1}$. Next, multiply by $\frac{1}{7}$ (which is the same as dividing by $\frac{7}{1}$): $\frac{9}{2} \times \frac{1}{7} = \frac{9}{14}$.

229. (c) It is 36 linear feet around the perimeter of the room (9×4). $36 - 17\frac{3}{4} = \frac{73}{4}$ or $18\frac{1}{4}$.

230. (c) Mixed numbers must be converted to fractions, and you must use the least common denominator of 8. $\frac{18}{8} + \frac{37}{8} + \frac{4}{8} = \frac{59}{8}$, which is $7\frac{3}{8}$ after it is reduced.

231. (d) Since 28 of the 35 slices have been eaten, there are $35 - 28 = 7$ slices left. This means $\frac{7}{35}$, or $\frac{1}{5}$ of the loaf is left.

1000 MATH PROBLEMS >>> Answers

232. (a) The common denominator is 24. $\frac{56}{24} - \frac{21}{24} = \frac{35}{24}$ or $1\frac{11}{24}$.

233. (c) Convert both the cost and the length to fractions. $\frac{3}{4} \times \frac{22}{3} = \frac{66}{12}$ or $5\frac{1}{2}$, which is Rs. 5.50.

234. (b) The total number of pages assigned is 80. $\frac{1}{6} \times 80 = \frac{80}{6}$ or $13\frac{1}{3}$.

235. (d) To subtract, convert to improper fractions, find a common denominator and subtract the numerators. $\frac{11}{2} - \frac{10}{3} = \frac{33}{6} - \frac{20}{6} = \frac{13}{6}$ or $2\frac{1}{6}$.

236. (c) To find the hourly wage, divide the total salary by the number of hours worked, or 331.01 divided by $39\frac{1}{2}$, converted to a decimal, which is 39.5, which equals 8.38.

237. (d) First, find how much of the cookie was eaten by adding the two fractions. After converting to the least common denominator the amount eaten is $\frac{9}{21} + \frac{7}{21} = \frac{16}{21}$. This means $\frac{5}{21}$ of the cookie is left.

238. (b) When subtracting mixed fractions, subtract the fractions first. Since 8 contains no fractions, convert to $7\frac{8}{8}$, then subtract, in this case, $\frac{8}{8} - \frac{5}{8} = \frac{3}{8}$. Then subtract the whole numbers, in this case $7 - 6 = 1$ (remember, 8 was converted to $7\frac{8}{8}$). Add the results, or $1\frac{3}{8}$.

239. (a) $\frac{3}{5}$ of 360 is figured as $\frac{3}{5} \times \frac{360}{1} = \frac{1080}{5}$ or 216.

240. (d) The common denominator of the fractions is 840. When added together, the fractions total $\frac{2039}{840}$ or $2\frac{427}{1000}$. The total of the whole numbers is 38. The total miles is $36 + 2\frac{427}{1000}$ or $38\frac{427}{1000}$.

SET 16

241. (c) The total number of items ordered is 36; the total received is 23. Therefore, Mona has received 23 of 36 items of $\frac{23}{36}$. $\frac{23}{36}$ cannot be reduced.

242. (b) To multiply fractions, convert to improper fractions, $\frac{31}{4} \times \frac{191}{5} = \frac{5921}{20}$ or Rs. 296.05.

243. (d) $\frac{1}{3}$ of $3\frac{1}{2}$ is expressed as $\frac{1}{3} \times 3\frac{1}{2}$, or $\frac{1}{3} \times \frac{7}{2} = \frac{7}{6}$ or $1\frac{1}{6}$.

1000 MATH PROBLEMS >>> Answers

244. (a) To divide $20\frac{1}{3}$ by $1\frac{1}{2}$, first convert to improper fractions, or $\frac{61}{3}$ divided by $\frac{3}{2}$. To divide, invert the second fraction and multiply. $\frac{61}{3} \times \frac{2}{3} = \frac{122}{9} = 13.55$ or approximately $13\frac{1}{2}$ recipes.

245. (c) There are two glasses out of eight left to drink, or $\frac{2}{8}$, which reduces to $\frac{1}{4}$.

246. (a) There will be 100 questions on the test. Veena has completed 25 of 100, or $\frac{1}{4}$ of the test.

247. (c) To figure the necessary number of gallons, divide the number of miles by the miles per gallon or $\frac{117}{2}$ divided by $\frac{43}{3} = \frac{351}{86}$ or $4\frac{7}{86}$.

248. (d) Uma still needs 18 squares, or $\frac{18}{168}$, which can be reduced to $\frac{3}{28}$.

249. (b) $14\frac{1}{2} \times 12\frac{1}{3}$, or $\frac{29}{2} \times \frac{37}{3} = \frac{1073}{6}$ or 178.83 square feet. Two gallons of paint will cover 180 square feet.

250. (b) The amount of meat loaf left is $\frac{2}{5} - (\frac{2}{5})(\frac{1}{4}) = \frac{2}{5} - \frac{2}{20}$. After finding the least common denominator this becomes $\frac{8}{20} - \frac{2}{20} = \frac{6}{20}$, which reduces to $\frac{3}{10}$.

251. (a) Meena has completed $\frac{16}{52}$ galleries, or $\frac{4}{13}$, of the total galleries.

252. (c) First, convert Rs. 12.35 to a fraction ($12\frac{35}{100}$), then convert to an improper fraction ($\frac{1235}{100}$) and reduce to $\frac{247}{20}$. $\frac{3}{4}$ of the work week is 30 hours. To multiply a whole number by a fraction, convert the whole number to a fraction, $\frac{30}{1}$. $\frac{247}{20} \times \frac{30}{1} = \frac{7410}{20}$ or $370\frac{1}{2}$ or Rs. 370.50.

253. (b) $\frac{1}{2}$ or $\frac{1}{4}$ is expressed as $\frac{1}{2} \times \frac{1}{4}$ or $\frac{1}{8}$.

254. (d) $\frac{1}{4} \times 2$ is expressed as $\frac{1}{4} \times \frac{2}{1} = \frac{2}{4}$, or $\frac{1}{2}$ teaspoon.

255. (a) The total area of the lawn is 810 square yards (30×27). There is $\frac{1}{3}$ of the yard left to mow. $\frac{1}{3} \times \frac{810}{1} = \frac{810}{3}$ or 270.

256. (d) The fraction of program time devoted to commercials is $\frac{6}{30}$, or $\frac{1}{5}$.

1000 MATH PROBLEMS >>> Answers

SET 17

257. (a) The formula is $10 = \frac{1}{12} \times x$, or 120.

258. (d) To determine the number of bags each hour, divide the total bags by the total hours, or $15\frac{1}{2}$ divided by 3, or $\frac{31}{2}$ divided by $\frac{3}{1}$, or $\frac{31}{2}$ times $\frac{1}{3}$, which is equal to $\frac{31}{6}$ or $5\frac{1}{6}$ bags per hour.

259. (c) To determine $\frac{1}{3}$ of $9\frac{3}{4}$, multiply $9\frac{3}{4}$ by $\frac{1}{3}$. After converting to fractions, this becomes $\frac{39}{4} \times \frac{1}{3} = \frac{39}{12}$, or $3\frac{1}{4}$ miles.

260. (a) This is a subtraction of mixed numbers problem. The common denominator is 8. Convert $\frac{1}{2}$ to $\frac{4}{8}$. Because $\frac{4}{8}$ is larger than $\frac{1}{8}$, you must borrow from the whole number 23. Then, subtract. $22\frac{9}{8} - 6\frac{4}{8} = 16\frac{5}{8}$.

261. (a) This is a subtraction problem. First, find the lowest common denominator, which is 12. $\frac{3}{4} = \frac{9}{12}$ and $\frac{2}{3} = \frac{8}{12}$. Then subtract: $\frac{9}{12} - \frac{8}{12} = \frac{1}{12}$.

262. (b) To solve this problem, you must first convert yards to inches. There are 36 inches in a yard. $36 \times 3\frac{1}{3} = 120$.

263. (b) This is a subtraction of fractions problem. The common denominator is 8. Convert $22\frac{1}{4}$ to $22\frac{2}{8}$. Because $\frac{5}{8}$ is larger than $\frac{2}{8}$, you must borrow. Then, subtract: $21\frac{10}{8} - 17\frac{5}{8} = 4\frac{5}{8}$.

264. (c) This is a simple subtraction problem with mixed numbers. First, find the lowest common denominator, which is 15. $\frac{4}{5} = \frac{12}{15}$ and $\frac{1}{3} = \frac{5}{15}$. Then, subtract. $20\frac{12}{15} - 3\frac{5}{15} = 17\frac{7}{15}$.

265. (d) First you must add the two fractions to determine what fraction of the total number of buses was in for maintenance and repair. The common denominator for $\frac{1}{6}$ and $\frac{1}{8}$ is 24, so $\frac{1}{6} + \frac{1}{8} = \frac{4}{24} + \frac{3}{24}$, or $\frac{7}{24}$. Next, divide 28 by $\frac{7}{24}$. $28 \div \frac{7}{24} = 96$.

266. (d) Mixed numbers must be converted to fractions, and you must use the least common denominator of 8: $\frac{18}{8} + \frac{37}{8} + \frac{4}{8} = \frac{59}{8}$, which is $7\frac{3}{8}$ after it is reduced.

267. (b) This is a problem involving addition of mixed numbers. First, find the common denominator, which is 6. Convert the fractions and add: $\frac{2}{6} + \frac{5}{6} + \frac{4}{6} = \frac{11}{6}$. Next, reduce the fraction: $\frac{11}{6} = 1\frac{5}{6}$. Finally, add the whole numbers and the mixed number: $2 + 1 + 2 + 1\frac{5}{6} = 6\frac{5}{6}$.

268. (b) Solve this problem with the following equation: $37\frac{1}{2}$ hours $- 26\frac{1}{4}$ hours $= 11\frac{1}{4}$ hours.

269. (a) Solve this problem with the following equations: $\frac{1}{2} + \frac{1}{2} + (5 \times \frac{1}{4}) = 1 + 1\frac{1}{4} = 2\frac{1}{4}$ miles.

270. (d) This is an addition of fractions problem. Add the top numbers of the fractions: $\frac{1}{3} + \frac{1}{3} = \frac{2}{3}$. Then add the whole number: $1 + \frac{2}{3} = 1\frac{2}{3}$.

271. (b) This is an addition problem with mixed numbers. First, add the fractions by finding the lowest common denominator; in this case, it is 40. $\frac{15}{40} + \frac{12}{40} + \frac{8}{40} = \frac{35}{40}$. Next, reduce the fraction: $\frac{35}{40} = \frac{7}{8}$. Then add the whole numbers: $7 + 6 + 5 = 18$. Finally, add the result: $18 + \frac{7}{8} = 18\frac{7}{8}$.

272. (d) This is an addition problem. To add mixed numbers, first add the fractions: To do this, find the lowest common denominator; in this case, it is 20. $\frac{3}{4} = \frac{15}{20}$ and $\frac{3}{5} = \frac{12}{20}$. Now add: $\frac{15}{20} + \frac{12}{20} = \frac{27}{20}$. Next, reduce: $\frac{27}{20} = 1\frac{7}{20}$. Finally, add the whole numbers to get the result: $16 + 2 + 1\frac{7}{20} = 19\frac{7}{20}$.

SET 18

273. (a) In this problem you must find the fraction. Mahima has completed 15 of the 25 minutes, of $\frac{15}{25}$. Reduce the fraction: $\frac{15}{25} = \frac{3}{5}$.

274. (c) This is a multiplication problem. To multiply a whole number by a mixed number, first convert the mixed number to a fraction: $1\frac{2}{5} = \frac{7}{5}$. Then, multiply $\frac{7}{5} \times \frac{3}{1} = \frac{21}{5}$. Now reduce. $\frac{21}{5} = 4\frac{1}{5}$.

275. (c) This is a division problem with mixed numbers. First, convert the mixed numbers to fractions: $33\frac{1}{2} = \frac{67}{2}$ and $5\frac{1}{4} = \frac{21}{4}$. Next, invert the second fraction and multiply: $\frac{67}{2} \times \frac{4}{21} = \frac{134}{21}$. Reduce to a mixed number: $\frac{134}{21} = 6\frac{8}{21}$. With this result, you know that only 6 glasses can be completely filled.

1000 MATH PROBLEMS >>> Answers

276. (d) This is a division problem. First, change the mixed number to a fraction: $3\frac{1}{2} = \frac{7}{2}$. Next, invert $\frac{1}{8}$ and multiply: $\frac{7}{2} \times \frac{8}{1} = 28$.

277. (b) This is a multiplication of fractions problem. Six minutes is $\frac{6}{60}$ of an hour, which is reduced to $\frac{1}{10}$; $2\frac{1}{2} = \frac{5}{2}$. Next, multiply: $\frac{1}{10} \times \frac{5}{2} = \frac{1}{4}$.

278. (d) First convert the $2\frac{1}{2}$ hours to minutes by multiplying $2\frac{1}{2} \times 60$ to get 150 minutes. Then divide the answer by 50, the number of questions: $150 \div 50 = 3$.

279. (b) Use 35 for C. $F = (\frac{9}{5} \times 35) + 32$. Therefore $F = 63 + 32$, or 95.

280. (b) Because the answer is a fraction, the best way to solve the problem is to convert the known to a fraction: $\frac{10}{45}$ of the cylinder is full. By dividing both the numerator and the denominator by 5, you can reduce the fraction to $\frac{2}{9}$.

281. (c) Solving this problem requires determining the circumference of the spool by multiplying $\frac{22}{7}$ by $3\frac{1}{2}$ ($\frac{7}{2}$). Divide the total (11) into 53. The answer is 4.8, so the hose will completely wrap only 4 times.

282. (c) First convert Fahrenheit to Centigrade using the formula given: $C = \frac{5}{9}(122 - 32)$; that is, $C = \frac{5}{9} \times 90$; so $C = 50$.

283. (b) To solve this problem, find the number of gallons of water missing from each tank ($\frac{1}{3}$ from Tank A, $\frac{3}{5}$ from Tank B), and then multiply by the number of gallons each tank holds when full (555 for Tank A; 680 for Tank B): $\frac{1}{3} \times 555 = 185$ gallons for Tank A; $\frac{3}{5} \times 680 = 408$ gallons for Tank B. Now add the number of gallons missing from both tanks to get the number of gallons needed to fill them: $185 + 408 = 593$.

284. (a) This is a multi-step problem. First, determine how many pieces of pizza have been eaten. Eight pieces ($\frac{8}{9}$) of the first pizza have been eaten; 6 pieces ($\frac{6}{9}$) of the second pizza have been eaten; 7 pieces ($\frac{7}{9}$) of the third pizza have been eaten. Next, add the eaten pieces. $8 + 6 + 7 = 21$. Since there are 27 pieces of pizza in all, 6 pieces are left, or $\frac{6}{27}$. Reduce the fraction. $\frac{6}{27} = \frac{2}{9}$.

285. (c) To solve the problem you have to first convert the total time to minutes (for the correct choice, (c), this is 105 minutes), then multiply by 4 (420 minutes), then convert the answer back to hours by dividing by 60 minutes to arrive at

1000 MATH PROBLEMS >>> Answers

the final answer (7 hours). Or you can multiply $1\frac{3}{4}$ hours by 4 to arrive at the same answer.

286. (a) In this problem, you must multiply a fraction by a whole number. $\frac{5}{6} \times \frac{5}{1} = \frac{25}{6}$. Reduce $\frac{25}{6} = 4\frac{1}{6}$.

287. (b) This is a two-step problem involving both addition and division. First add the two mixed numbers to find out how many ounces of jelly beans there are in all: $10\frac{1}{4} + 9\frac{1}{8} = 19\frac{3}{8}$. Convert the result to a fraction: $19\frac{3}{8} = \frac{155}{8}$. Next, to divide, invert the whole number and multiply: $\frac{155}{8} \times \frac{1}{5} = \frac{31}{8}$. Reduce. $\frac{31}{8} = 3\frac{7}{8}$.

288. (b) If half the students are female, then you would expect half of the out-of-state students to be female. One half of $\frac{1}{12}$ equals $\frac{1}{2} \times \frac{1}{12}$, or $\frac{1}{24}$.

SET 19

289. (b) There are 60 minutes in an hour. Multiply $60 \times 7\frac{1}{6}$ by multiplying $60 \times 7 = 420$ and $60 \times \frac{1}{6} = 10$. Then add $420 + 10$ to get 430 minutes.

290. (a) To solve this problem, you must convert $3\frac{1}{2}$ to $\frac{7}{2}$ and then divide $\frac{7}{2}$ by $\frac{1}{4}$. The answer, $\frac{28}{2}$, is then reduced to the number 14.

291. (c) The simplest way to add these three numbers is to first add the whole numbers, then add the fractions: $3 + 2 + 4 = 9$. Then, $\frac{1}{2} + \frac{3}{4} = \frac{2}{4} + \frac{3}{4} = \frac{5}{4}$, or $1\frac{1}{4}$. Then, $9 + 1\frac{1}{4} = 10\frac{1}{4}$.

292. (c) This is a multiplication problem involving a whole number and a mixed number. There are 16 ounces in a pound so you must multiply $16 \times 9\frac{1}{2}$. First, change the whole number and the mixed number to fractions: $\frac{16}{1} \times \frac{19}{2} = \frac{304}{2}$. Converting: $\frac{304}{2} = 152$ ounces.

293. (b) In this problem you must find the fraction: $\frac{7,500}{25,000}$. Next, reduce the fraction. The easiest way to reduce is to first eliminate zeros in the numerator and the denominator: $\frac{7,500}{25,000} = \frac{75}{250}$. Then, further reduce the fraction. $\frac{75}{250} = \frac{3}{10}$.

294. (d) If 2 of 5 cars are foreign, 3 of 5 are domestic. $\frac{3}{5} \times 60$ cars = 36 cars.

1000 MATH PROBLEMS >>> Answers

295. (b) $2\frac{1}{2}$ is equal to 2.5. $1\frac{1}{4}$ is equal to 1.25. 2.5×1.25 is equal to 3.12 or $3\frac{1}{8}$.

296. (b) Because the answer is a fraction, the best way to solve the problem is to convert the known to a fraction: $\frac{10}{45}$ of the test tube is full. By dividing both the numerator and the denominator by 5, you can reduce the fraction to $\frac{2}{9}$.

297. (b) Use the formula beginning with the operation in parentheses: 98 minus 32 equals 66. Then multiply 66 by $\frac{5}{9}$, first multiplying 66 by 5 to get 330. $330 \div 9 = 36.67$, which is rounded up to 36.7.

298. (d) Use the formula provided: $\frac{9}{5}(40) + 32 = 72 + 32 = 104$.

299. (d) This is a multiplication of fractions problem: $\frac{1}{3} \times \frac{210}{1} = 70$.

300. (d) This is a two-step problem involving multiplication and simple subtraction. First, determine the amount of sand contained in the 4 trucks. $\frac{3}{4} \times \frac{4}{1} = \frac{12}{4}$. Reduce: $\frac{12}{4} = 3$. Then, subtract. $3 - 2\frac{5}{6} = \frac{1}{6}$. There is $\frac{1}{6}$ ton more than is needed.

301. (b) To solve the problem, you must first find the common denominator, in this instance, 24. Then the fractions must be converted: $\frac{1}{8} = \frac{3}{24}$; $\frac{1}{6} = \frac{4}{24}$; $\frac{3}{4} = \frac{18}{24}$. Add the values for first and second layers together: $\frac{3}{24} + \frac{4}{24} = \frac{7}{24}$, then subtract the sum from the total thickness ($\frac{18}{24}$): $\frac{18}{24} - \frac{7}{24} = \frac{11}{24}$.

302. (c) The recipe for 16 brownies calls for $\frac{2}{3}$ cup butter. An additional $\frac{1}{3}$ cup would make 8 more brownies, for a total of 24 brownies.

303. (a) The recipe is for 16 brownies. Half of that, 8, would reduce the ingredients by half. Half of $1\frac{1}{2}$ cups of sugar is $\frac{3}{4}$ cup.

304. (c) In this problem you must find the fraction. $\frac{6}{10}$ of the cake has been eaten, so $\frac{4}{10}$ of the cake is left. Now reduce the fraction: $\frac{4}{10} = \frac{2}{5}$.

SET 20

305. (a) Again, rounding to close numbers helps. This is approximately $100,000 \div 500,000 = 0.20$ or $\frac{1}{5}$.

306. (b) Yellow beans + orange beans = 12. There are 30 total beans. $\frac{12}{30}$ is reduced to $\frac{2}{5}$.

1000 MATH PROBLEMS >>> Answers

307. (c) In this problem you must multiply a mixed number by a whole number. First, rewrite the whole number. First, rewrite the whole number as a fraction. $5 = \frac{5}{1}$. Then rewrite the mixed number as a fraction: $6\frac{1}{2} = \frac{13}{2}$. Then multiply: $\frac{5}{1} \times \frac{13}{2} = \frac{65}{2}$. Finally, convert to a mixed number. $\frac{65}{2} = 32\frac{1}{2}$.

308. (c) This is a division of fractions problem. First, change the whole number to a fraction: $6 = \frac{6}{1}$. Then invert the second fracton and multiply : $\frac{6}{1} \times \frac{4}{1} = 24$.

309. (d) In this problem you must multiply a fraction by a whole number. First, rewrite the whole number as a fraction: $8 = \frac{8}{1}$. Next, multiply $\frac{8}{1} \times \frac{4}{5} = \frac{32}{5}$. Finally, convert to a mixed number: $\frac{32}{5} = 6\frac{2}{5}$.

310. (c) This is a multiplication with mixed numbers problem. First, change both mixed numbers to fractions: $3\frac{1}{4} = \frac{13}{4}$; $1\frac{2}{3} = \frac{5}{3}$. Next, multiply the fractions: $\frac{13}{4} \times \frac{5}{3} = \frac{65}{12}$. Finally, change the result to a mixed number: $\frac{65}{12} = 5\frac{5}{12}$.

311. (b) This is a division problem. First, change the mixed number to a fraction: $1\frac{1}{8} = \frac{9}{8}$. Invert the whole number 3 and multiply: $\frac{9}{8} \times \frac{1}{3} = \frac{3}{8}$.

312. (b) This is a basic addition problem. First, change the fractions to decimals. Then, you might think of the problem this way: For Arun to get to Rahul's house, he must go 2.5 miles west to the Sunnydale Mall, then on west 1.5 miles to the QuikMart, and then on 4.5 miles to Rahul's house. So: 2.5 + 1.5 + 4.5 = 8.5 miles.

313. (a) To find the area of the floor in square feet, multiply the length by the width, or $9\frac{3}{4} \times 8\frac{1}{3}$. To multiply mixed numbers first convert to improper fractions, or $\frac{39}{4} \times \frac{25}{3} = \frac{975}{12}$ or $81\frac{1}{4}$.

314. (b) There are 12 inches in a foot. Change the mixed number to a decimal: 4.5. Now multiply: 4.5 × 12 = 54.

315. (a) First, change the mixed numbers to decimals: $7\frac{1}{2} = 7.5$ and $4\frac{1}{4} = 4.25$. Now subtract: 7.5 − 4.25 = 3.25. Now change the decimal back to a fraction: $3.25 = 3\frac{1}{4}$.

316. (c) First, find the least common denominator of the fractions, which is 6. Then add: $9\frac{2}{6} + 8\frac{5}{6} = 17\frac{7}{6}$, or $18\frac{1}{6}$.

1000 MATH PROBLEMS >>> Answers

317. (d) First add the weight of Rani's triplets: $4\frac{3}{8} + 3\frac{5}{6} + 4\frac{7}{8}$, or (after finding the least common denominator) $4\frac{9}{24} + 3\frac{20}{24} + 4\frac{21}{24} = 11\frac{50}{24}$, or $13\frac{2}{24}$, or $13\frac{1}{12}$. Now find the weight of Reema's twins: $7\frac{2}{6} + 9\frac{3}{10}$, or (after finding the least common denominator) $7\frac{10}{30} + 9\frac{9}{30} = 16\frac{19}{30}$. Now subtract: $16\frac{19}{30} - 13\frac{1}{12} = 16\frac{38}{60} - 13\frac{5}{60} = 3\frac{33}{60}$. So Reema's twins outweigh Rani's triplets by $3\frac{33}{60}$. (No further reduction of the fraction is possible.)

318. (b) There are 16 ounces in a pound, so Mohan ate $\frac{5}{8} \times 16$ ounces. Multiply: $\frac{5}{8} \times \frac{16}{1} = \frac{80}{8}$ ounces, or 10 ounces.

319. (a) Approach this problem by placing the total amount of Beena's gift over the amount Beena spent: $\frac{4500}{450} = 0.1$ or $\frac{1}{10}$.

320. (c) Change the information into a fraction: $\frac{2}{8}$. Now, reduce the fraction: $\frac{1}{4}$.

SECTION 3: DECIMALS

SET 21

321. (d) The hundredth is the second digit to the right of the decimal point. Because the third decimal is 6, the second is rounded up to 4.

322. (d) The farther to the right the digits go, the smaller the number.

323. (c) Common errors include (a), subtracting instead of adding; or (d), not lining up the decimal points correctly.

324. (c) If you line up the decimal points properly, you don't even have to do the subtraction to see that none of the other answers is even close to the correct value.

325. (a) To work this kind of division problem, you have to move each decimal point to the right and then bring the point straight up into the answer. If you got item (c) you divided 0·8758 by 2.9.

326. (b) 195·6 ÷ 7.2 yields a repeating decimal, 27·1666666..., which, rounded up to the nearest hundredth, is 27.17.

327. (c) The other answers were subtracted without aligning the decimal points.

328. (a) The correct answer is 31.75. Not lining up the decimal points when multiplying is the most common error in this type of problem.

329. (d) This is a basic subtraction problem with decimals. The correct answer is 1.734.

330. (b) The correct answer has only two decimal places: 96.32.

331. (c) Since there are two decimal places in each number you are multiplying, you need a total of four decimal places in the answer, 0.2646.

332. (c) This is a mixed decimal. The whole number, 6, is to the left of the decimal point. The hundredths place is the second digit to the right of the decimal point.

333. (d) The hundredths place is two digits to the right of the decimal point. The 9 is in the tenths place; the 0 is in the thousandths place; the 2 is in the ten-thousandths place.

334. (a) 0.7 is the correct answer. Choice (b) is rounded up instead of down; choice (c) is rounded to the nearest hundredths place; and choice (d) is rounded to the nearest whole number.

1000 MATH PROBLEMS >>> Answers

335. (a) In (b), the 9 is in the hundredths place. In (c), it is in the tenths place; in (d), the ten-thousandths place.

336. (c) The correct answer is 1.8.

SET 22

337. (a) You have to be careful to align the decimal points.

338. (b) You need four digits to the right of the decimal point.

339. (c) It is important to keep the decimal values straight. Divide as usual, and then bring the decimal point straight up into the answer in order to get 39.4.

340. (b) Think of 145.29 as 145.290, and then line up the decimal points and add the numbers.

341. (d) You can rule out (a) and (c) because the 7 in the thousandths place indicates subtraction rather than addition. You would get answer (b) only if you added 8 and 9 incorrectly.

342. (c) This is a simple subtraction problem, as long as the decimals are lined up correctly.

343. (c) The solution to this problem lies in knowing that 100^2 is equal to 100×100, or 10,000. Next, you must multiply $10,000 \times 2.75$ to arrive at 27,500.

344. (d) This is a simple division problem with decimals.

345. (c) This is a simple multiplication problem with decimals.

346. (c) 10 times 10 times 10 is 1000. 1000 times 7.25 is 7250.

347. (d) To multiply two numbers expressed in scientific notation, multiply the non-exponential terms (4.1 and 3.8) in the usual way. Then the exponential terms (10^{-2} and 10^4) are multiplied by adding their exponents. So $(4.1 \times 10^{-2})(3.8 \times 10^4) = (4.1 \times 3.8)(10^{-2} \times 10^4) = (15.58)(10^{-2+4}) = (15.58)(10^2) = 15.58 \times 10^2$. In order to express this result in scientific notation, you must move the decimal point one place to the left and add one to the exponent, resulting in 1.558×10^3.

348. (b) To divide two numbers in scientific notation, you must divide the non-exponential terms (6.5 and 3.25) in the usual way, then divide the exponential terms (10^{-6} and 10^{-3}) by subtracting the exponent of the bottom term from the exponent of the top term, so that you get

$$\frac{(6.5 \times 10^{-6})}{(3.25 \times 10^{-3})} = \frac{6.5}{3.25} \times \frac{10^{-6}}{10^{-3}} = 2 \times 10^{-6 - (-3)} = 2 \times 10^{-6 + 3} = 2 \times 10^{-3}.$$

349. (d) To find the answer do the following equation: $11 \times 0.032 = 0.352$.

350. (b) The last digit has to be a 3, which rules out (c). You can rule out (a) and (d) because of their place value.

351. (a) −0.15 is less than −0.02, the smallest number in the range.

1000 MATH PROBLEMS >>> *Answers*

SET 23

352. (d) Four digits to the right of the decimal point is the ten-thousandths place.

353. (a) Choice **(b)** can easily be ruled out because the 0 in the tenths place is less than the 7 in all the other choices. Choices **(c)** and **(d)** can be ruled out because the digits in the thousandths place are less than 2.

354. (a) Because there are zeros in both the tenths and hundredths places the other choices are all greater than choice **(a)**.

355. (b) The second zero to the right of the decimal point is in the hundredths place. Because the next digit, 9, is greater than 5 round up from 0 to 1.

356. (b) First arrange the numbers in a column so that the decimal points are aligned. Then add. The sum 45.726 is then rounded to 45.7 because the 2 in the hundredths place is less than 5.

357. (c) The correct answer is 91.9857.

358. (b) The correct answer is 1.48.

359. (c) The equation to be used is (70T + 60T) = 325.75, or 130T = 325.75; T = 2.5.

360. (b) Rs. 4.20 = 75% of full price; therefore, the full price is Rs. 5.60.

361. (c) To add, write the first number as 1.50 and add to 1.35 = 2.85.

362. (a) To subtract, write the first number as 17.80 and subtract 14.33. The answer is 3.47.

363. (b) The total of the three numbers is 34.47.

364. (d) Add Rs. 157.50 to Rs. 679.80 and then subtract Rs. 275.80. The answer is Rs. 561.5.

365. (c) First, add the cost of the wallpaper and carpet, which is Rs. 701.99. Subtract that from Rs. 768.56. The difference is Rs. 66.57.

366. (a) 100 kgs is four times 25 kgs, so the cost is Rs. 3.59 times 4, or Rs. 14.36.

367. (d) 63 divided by 5 equals 12.6.

368. (b) Rs. 37.27 multiplied by six equals Rs. 223.62.

SET 24

369. (c) 5.96 divided by 4 equals 1.49.

370. (a) There are twelve inches in a foot. 2.54 multiplied by 12 is 30.48.

371. (c) In the first two hours, they are working together at the rate of 1 aisle per hour for a total of two aisles. In the next two hours, they work separately. Meena works for two hours at 1.5 hours per aisle. 2 ÷ 1.5 = 1.33 aisles. Jona works for two hours at two hours per aisle. This is one aisle. The total is 4.33 aisles.

1000 MATH PROBLEMS >>> Answers

372. (c) First change the fraction to a decimal: $\frac{1}{8} = 1 \div 8 = 0.125$. Now multiply that by Hari's hourly wage in order to get Pran's hourly wage: $0.125 \times$ Rs. 19.50 = Rs. 2.4375 (rounded). Now multiply Pran's hourly wage by 8 hours. Rs. $2.4375 \times 8 = 19.50$.

373. (b) 45 minutes is equal to $\frac{3}{4}$ of an hour, so Reema will only make $\frac{3}{4}$ of her usual fee. Change $\frac{3}{4}$ to a decimal: 0.75. Now multiply: $10 \times 0.75 = 7.5$. Reema will make Rs. 7.50 today.

374. (c) Change the fraction to a decimal: 0.25 (which is $\frac{1}{4}$ as fast or 25% of Heera's time). Now multiply: $35.25 \times 0.25 = 8.8125$, which can be rounded to 8.81.

375. (c) This problem is done by dividing: $1.5 \div 600 = 0.0025$ inches.

376. (b) We are looking for D = the distance to the ceiling in feet. The time the ant starts its journey (6:07 a.m.) is irrelevant to the problem. Since the ant is travelling at a uniform rate, we can use the formula Distance = Rate × Time, or D = RT, where $R = \frac{48 \text{ ft}}{\text{hr}} = 0.8 \frac{\text{ft}}{\text{min}}$ and T = 10 min. Therefore D = 0.8(10) = 8 ft.

377. (b) Kiran saves Rs. 60 + Rs 130 + Rs 70 which equals Rs 260. In January, her employer contributes Rs. $60 \times 0.1 =$ Rs. 6, and in April, her employer contributes Rs. $70 \times 0.1 =$ Rs. 7. In March, her employer contributes only Rs. 10 (not Rs. 13), because Rs. 10 is the maximum employer contribution. The total in savings then is Rs. 260 + Rs. 6 + Rs. 7 + Rs 10 = Rs 283. (If you chose option (c), you forgot that the employer's contribution was a *maximum* of Rs. 10.)

378. (a) This is a subtraction problem. Align the decimals and subtract. $428.99 - 399.99 = 28.99$.

379. (c) This is a multiplication problem with decimals. $2.5 \times 2{,}000 = 5{,}000$.

380. (c) This is a division problem: $304.15 \div 38.5$. Because there is one decimal point in 38.5, move the decimal point one place in both numbers. $\frac{3041.5}{385} = 7.9$.

381. (c) This is a multiplication problem. To quickly multiply a number by 1,000, move the decimal point three digits to the right—one digit for each zero. In this situation, because there are only two decimal places, add a zero.

382. (d) This is a division problem. Divide 12.9 by 2 to get 6.45, then add both numbers. $12.90 + 6.45 = 19.35$.

383. (a) This is a division problem. Because there are two decimal points in 1.25, move the decimal point two places in both numbers. $\frac{2240}{125} = 17.92$.

384. (b) This is a simple subtraction problem. Be sure to align the decimal points. $99.0 - 97.2 = 1.8$.

1000 MATH PROBLEMS >>> Answers

SET 25

385. (b) Both addition and subtraction are required to solve this problem. First add the amounts of the three purchases together: 12.90 + 0.45 + 0.88 = 14.23. Next, subtract this amount from 40.00. 40.00 − 14.23 = 25.77.

386. (d) This problem requires both multiplication and addition. First, multiply 2.12 by 1.5 to find the price of the cheddar cheese: 2.12 × 1.5 = 3.18. Then add. 2.12 + 2.34 = 5.52.

387. (b) This problem requires both addition and subtraction. First, add the three lengths of string: 5.8 + 3.2 + 4.4 = 13.4. Then subtract the answer from 100, making sure to align the decimal points: 100.0 − 13.4 = 86.6.

388. (c) This is a simple division problem. $\frac{2.00}{3}$ = 0.666. Because 6 is higher than 5, round upto 7.

389. (a) This is a simple addition problem. Line up the decimals in a column so that the decimal points are aligned. 9.4 + 18.9 + 22.7 = 51.

390. (b) This is a multiplication problem with decimals. Manish spends Rs. 1.10 each way and makes 10 trips each week: 1.10 × 10 = 11.00.

391. (d) This is a multiplication problem. 0.39 × 0.25 = 0.0975. Be sure to count off a total of four decimal points in your answer.

392. (c) 5.133×10^{-6} = 5.133 × 0.000001 = 0.000005133. This is the same as simply moving the decimal point to the left 6 places.

393. (a) Add the four numbers. The answer is Rs. 23.96.

394. (b) Add all of the money gifts as well as what Hari earned. The total is Rs. 4772.56. Subtract this number from the cost of the car. The remainder is Rs. 227.39.

395. (c) Subtract 2.75 from 10 by adding a decimal and zeros to the 10. 10.00 − 2.75 = 7.25. Don't forget to line up the decimals.

396. (d) Remember to ignore the decimals when multiplying 8245 × 92 = 758540. Then, total the number of decimal points in the two numbers you multiplied (four) and place the decimal point four places from the right in the answer: 75.8540.

397. (b) Divide 517.6 by 23.92. Don't forget to move both decimals two spaces to the right (add a zero to 517.6), move the decimal up directly and divide. The answer is 21.6387. Round up to 21.64.

398. (d) To find the difference, subtract 22.59 from 23.7. To keep the decimal placement clear, remember to add a 0 to 23.7.

399. (a) Add the three amounts, adding zeros where necessary. The total is 11.546, rounded down to 11.5 kgs.

400. (c) This is a basic multiplication problem: 25.56 × 5 = 127.80.

1000 MATH PROBLEMS >>> *Answers*

SET 26

401. (a) Add all the candy Indra distributed or consumed and subtract that number from 7.5. The result is 5.44 kgs.

402. (c) Add all the distances together. The sum is 16.32.

403. (b) 20.34 divided by 4.75 = 4.282. Kiran can cover four chairs.

404. (d) The total balance before expenditures is Rs. 2045.38. The total expenditure is Rs. 893.14. Subtracted, the total is Rs. 1152.24.

405. (a) Multiply the length by the width. The answer is 8.97 square feet.

406. (b) Five kgs of flour divided by .75 equals 6.6666 ... Manish can make 6 cakes.

407. (d) Add all the distances and divide by the number of days; 4.38 divided by 4 = 1.095.

408. (c) Add the eaten amounts and subtract from the total. 12.600 − 6.655 = 5.945 kgs of clams are left. (If you chose option (d), you forgot the last step of the problem.)

409. (a) To find the difference, subtract 4.892 from 13.8. The difference is 8.908 million.

410. (c) Rs. 75.00 divided by Rs. 3.78 equals 19.8412. Round up to 20.

411. (b) Subtract the hours actually worked from the hours usually worked; the number of hours is 4.75.

412. (d) Add the three amounts; the total is 2.83 acres.

413. (d) To find the area, multiply the length by the width. The area is 432.9 square feet.

414. (a) The total of ads and previews is 11.3 minutes. Two hours is 120 minutes. 120 − 11.3 = 108.7.

415. (c) 10 metres divided by .65 equals 15.384615 (repeating). Tony can make 15 hats.

416. (b) Add the four amounts. The total is 28.13 hours.

SET 27

417. (a) The formula for determining the amount of fence needed is 2(78.45 + 65.89). The total amount needed is 288.68.

418. (c) 8.6 million minus 7.9 million equals 0.7 million.

419. (b) Rs. 1.13 multiplied by 100 equals Rs. 113.00. Remember, a shortcut for multiplying fractions by 10, 100, 1000, etc. is to simply move the decimal to the right one space for each zero.

420. (d) The four distances added together equal 5.39 miles.

421. (b) Fifteen minutes is $\frac{1}{4}$, or .25, of an hour. .25 of 46.75 is .25 × 46.75 = 11.6875.

422. (a) Multiply 46.75 by 3.80, which equals 177.65.

1000 MATH PROBLEMS >>> Answers

423. (d) The ratio of pirates to clergy is 3 : 4, or $\frac{3}{4}$, or 0.75. 88 × 0.75 = 66 pirate ancestors.

424. (a) This is a division problem with decimals. 4446 ÷ 11.70 = 380.

425. (b) You must divide two decimals: 20.32 ÷ 2.54. First, move each number over two decimal places: 2032 ÷ 254 = 8.

426. (c) This is a four-step problem. First, determine how much she earns in one 8 hour day: 8 × Rs. 12.50 = Rs. 100. Next, subtract Rs. 100 from Rs. 137.50 to find how much overtime she earned: Rs 137.50 – Rs. 100 = Rs. 37.50. Next, to find out how much her hourly overtime pay is, multiply 1.5 × 12.50, which is 18.75. To find out how many overtime hours she worked, divide: 37.50 ÷ 18.75 = 2. Add these hours to her regular 8 hours for a total of 10 hours.

427. (b) Solving this problem requires converting 15 minutes to 0.25 hour, which is the time, then using the formula: 62 mph × 0.25 hour = 15.5 miles.

428. (d) Distance travelled is equal to velocity (or speed) multiplied by time. Therefore, $3.00 \times (10^8) \frac{\text{meters}}{\text{second}} \times 2000$ seconds = 6.00×10^{11} metres.

429. (a) First it is necessary to convert centimetres to inches. To do this for choice (a), multiply 100 cm (1 metre) by 0.39 inches, yielding 39 inches. For choice b, 1 yard is 36 inches. For choice (d), multiply 85 cm by 0.39 inches, yielding 33.15 inches. Choice (a), 39 inches, is the longest.

430. (a) To solve this problem, divide the number of pounds (168) by the number of kilograms in a pound (2.2): 168 ÷ 2.2 = 76.36. Now round to the nearest unit, which is 76.

431. (b) This is a multiplication problem. Rs. 1.25 times 40 is Rs. 50.00

432. (b) You simply add all the numbers together to solve this problem.

SET 28

433. (c) The answer is arrived at by first dividing 175 by 45. Since the answer is 3.89, not a whole number, the firefighter needs 4 sections of hose. Three sections of hose would be too short.

434. (c) The answer to this question lies in knowing that there are four quarts in a 5-gallon container. Since 1 litre is a little more than 1 quart, there will be fewer litres than quarts in the container; 1 quart = 0.94 litre; (20)(0.94) = 18.8 litres in the container. Round up to 19 litres.

435. (c) To solve the problem, take the weight of one gallon of water (8.35) and multiply it by the number of gallons (25): 8.35 × 25 = 208.7. Now round to the nearest unit, which is 209.

436. (a) Because there are three at Rs. 0.99 and 2 at Rs. 3.49, the sum of the two numbers minus Rs. 3.49 will give the cost.

1000 MATH PROBLEMS >>> Answers

437. (d) 2200(0.07) equals Rs. 154. 1400(0.04) equals Rs. 56. 3100(0.08) equals Rs. 248. 900(0.03) equals Rs. 27. Therefore, Rs. 154 + Rs 56 + Rs 248 + Rs 27 = Rs 485.

438. (a) To solve the problem, multiply 3.5 kgs by 7, the number of days in one week.

439. (b) The solution is simply the ratio of the rates of work, which is 15.25:12.5, or $\frac{15.25}{12.5}$ or 1.22. (To check your work multiply: 12.5 units × 1.22 hours = 15.25).

440. (d) Rs. 12.50 per hour × 8.5 hours per day × 5 days per week is Rs. 531.25.

441. (d) 3 inches every 2 hours = 1.5 inches per hour × 5 hours = 7.5 inches.

442. (d) This can be most quickly and easily solved by estimating; that is, 2 hamburgers at Rs. 20.95 = Rs. 41.90. Rs. 41.90 + 1 Dosa at 30.35 = Rs 72.25. 2 chicken sandwiches at Rs. 30.95 = Rs. 61.90. Then: Rs. 61.90 + 1 grilled cheese at Rs. 10.95 = Rs. 72.85. Rs. 72.25 + Rs. 72.85 = Rs. 145.10.

443. (c) This is a simple multiplication problem, which is solved by multiplying 35 times 8.2 for a total of 287.

444. (d) This problem is solved by dividing 60 (the time) by 0.75 (the rate), which gives 80 words.

445. (c) To find the answer, work this equation: Rs. (12.24 − Rs. 12.08) × 2 = Rs. 0.32.

446. (d) First find the total price of the pencils: 24 pencils × Rs. 0.5 = Rs. 12/-. Then find the total price of the paper: 3.5 reams × Rs. 75/- per ream = Rs 262.50. Next, add the two totals together: Rs. 1.20 + 262.5 = Rs. 274.5.

447. (c) This is a two-step multiplication problem. To find out how many heartbeats there would be in one hour, you must multiply 72 by 60 minutes, and then multiply this result, 4,320, by 6.5 hours.

448. (d) This is a multiplication problem. Multiply 6.5 by 1.5. Because there are a total of two decimal digits, count off two places from the right before placing the decimal point.

SET 29

449. (c) This is a simple addition problem. Add 1.6 and 1.5, keeping the decimal points aligned: 1.6 + 1.5 = 3.1.

450. (b) This is a division problem: 25.8 ÷ 3 = 8.6. Move the decimal point straight up into the quotient.

451. (c) You arrive at this answer by knowing that 254 is one hundred times 2.54. To multiply by 100, move the decimal point two digits to the right.

452. (a) This is an addition problem. Add the three numbers together, making sure that the decimal points are aligned.

453. (d) This is a multiplication problem. 35.2 × 71 = 2499.2. There is only one decimal point, so you will count off only one place from the right.

1000 MATH PROBLEMS >>> *Answers*

454. (b) This is a multiplication problem. First multiply Rs. 279 by 89. Then, because there are four decimal places, count off four places from the right. Your answer should be 73.6831. Because the 3 in the thousandths place is less than 5, round to 73.68.

455. (a) This is a basic subtraction problem. Line up the decimals and subtract: 2354.82 − 867.59 = 1487.23.

456. (b) This is a division problem. 13.5 ÷ 4 = 3.375. Move the decimal straight up into the quotient.

457. (b) This is a simple subtraction problem. Line up the decimals and subtract. 46.1 − 40.6 = 5.5.

458. (d) This is an addition problem. To add these three decimals, line them up in a column so that their decimal points are aligned. (52.50 + 47.99 + 49.32 = 149.81). Move the decimal point directly down into the answer.

459. (c) This is a basic addition problem. Be sure to align the decimal points before you add 68.8 + 0.6 = 69.4.

460. (b) This is an addition problem. Arrange the numbers in a column so that the decimal points are aligned. 2.345 + 0.0005 = 2.3455.

461. (c) This is a multiplication problem. Multiply 3.25 times 1.06. Be sure to count 4 decimal places from the right. 3.25 × 1.06 = 3.445.

462. (c) This is an addition problem. Arrange the three numbers in a column so that the decimal points are aligned: 139.50 + 57.00 + 48.90 = 245.4.

463. (c) This is an addition problem with decimals. Add the four numbers together to arrive at the answer, which is Rs. 1772.10.

464. (d) This is a two-step multiplication problem. First, find out how long it would take for both Bunny and Sunny to do the job. 0.67 × 5 = 3.35. Then, multiply your answer by 2 because it will take Bunny twice as long to complete the job alone. 3.35 × 2 = 6.7.

SET 30

465. (a) This is a simple subtraction problem. Align the decimal points and subtract: 91,222.30 − 84,493.26 = 6,729.04.

466. (a) This is a division problem. Because there are two decimal digits 3.75, move the decimal point two places to the right in both numbers. This means you must take a zero on to the end of 2328. Then divide: 23,280 ÷ 375 = 62.08.

467. (a) This is a three-step problem that involves multiplication, addition, and subtraction. First, to determine the cost of the Dal, multiply 31.60 by 4. 31.60 × 4 = 126.40. Then add the price of both the Dal and the Rice. 126.40 + 128.40 = 254.80. Finally, subtract to find out how much money is left: 1000.00 − 254.80 = 745.20.

1000 MATH PROBLEMS >>> Answers

468. (c) This is a two-step problem involving multiplication and division. First, determine the length of the pipes in inches by multiplying: 15.4 × 3 = 46.2. Next, divide to determine the length in feet. 46.2 ÷ 12 = 3.85. Because there are no decimal points in 12, you can move the decimal point in 46.2 straight up into the quotient.

469. (c) This is a multiplication problem. Be sure to count four decimal places from the right in your answer: 28.571 × 12.1 = 345.7091.

470. (b) This is an addition problem. Arrange the three numbers in a column and be sure that the decimal points are aligned. Add: 0.923 + 0.029 + 0.1153 = 1.0673.

471. (c) This is a two-step problem involving both addition and division. First, arrange the three numbers in a column, keeping the decimal points aligned. Add: 113.9 + 106.7 + 122 = 342.6. Next, divide your answer by 3: 342.6 ÷ 3 = 114.2.

472. (b) This is an addition problem. Be sure the decimal points are aligned before you add. 0.724 + 0.0076 = 0.7316.

473. (d) This problem involves two steps: addition and subtraction. Add to determine the amount of money Mahesh has: 20.00 + 5.00 + 1.29 = 26.29. Then, subtract the amount of the ice cream: 26.89 − 4.89 = 21.40.

474. (c) This is a two-step problem. First, multiply to determine how many kilograms of beef were contained in the 8 packages. 0.75 × 8 = 6. Then add. 6 + 0.04 = 6.04.

475. (d) This is a two-step multiplication problem. First multiply. 5 × 2 = 10, which is the number of trips Jenny drives to get to work and back. Then multiply 19.85 by 10 by simply moving the decimal one place to the right.

476. (a) This is a simple subtraction problem. 42.09 − 6.25 = 35.84.

477. (c) This is a three-step problem. First, multiply to determine the amount Arun earned for the first 40 hours he worked. 40 × 8.3 = 332. Next, multiply to determine his hourly wage for his overtime hours: 1.5 × 4 × 8.30 = 49.8. Finally, add the two amounts. 332 + 49.8 = 381.8.

478. (c) This is a two-step problem involving both addition and subtraction. First add, 93.6 + 0.8 = 94.4. Then, subtract. 94.4 − 11.6 = 82.8.

479. (a) This is a two-step multiplication problem. First, multiply to find out how many weeks there are in 6 months: 6 × 4.3 = 25.8. Then, multiply to find out how much is saved: Rs. 40 × 25.8 = Rs. 1,032.

480. (a) This is a subtraction problem. Be sure to align the decimal points. 6.32 − 6.099 = 0.221.

SECTION 4 : PERCENTAGES

SET 31

481. (c) A decimal point is always understood to precede the percent sign. To change a percent to a decimal, remove the percent sign and move the decimal point two places to the left. 2% becomes 2.0 becomes 0.02. (It is best to place a zero before the decimal in order to avoid confusion.)

482. (a) The percent sign has been dropped and the decimal moved two places to the left.

483. (b) Again, the percent sign has been dropped and the decimal point moved two places to the left.

484. (c) Again, the percent sign has been dropped and the decimal point moved two places to the left.

485. (c) Remember, the percentage is always placed over 100: $\frac{400}{100} = 4$.

486. (b) Convert the decimal to a fraction: $\frac{2}{100} = 2 \div 100$, or 2.0. Now add the percent sign to get 2.0%.

487. (b) Convert the mixed number to a decimal: 6.25%.

488. (a) Again, the decimal point has been moved two places to the left.

489. (c) Change the fraction to a decimal—keep the percent sign where it is. $\frac{1}{4}\% = 0.25\%$.

490. (a) Change the fraction to a decimal, then the decimal to a percent: $\frac{1}{4} = 0.25 = 25\%$. (To change a decimal to a percent, move the decimal point two places to the right and add the percent sign.)

491. (a) Convert from a decimal to a percentage by multiplying by 1 in the form of $\frac{100}{100}$. Thus $0.97 \times \frac{100}{100} = \frac{97}{100}$ or 97%.

492. (b) $10\% = \frac{10}{100}$ (that is, 10 per 100), or $\frac{1}{10}$.

493. (b) First, change the percentage to a decimal by placing $\frac{350}{100}$, then convert to a mixed number by dividing: $350 \div 100 = 3.5$ or $3\frac{1}{2}$.

494. (d) $24\% = \frac{24}{100}$; reduced this is $\frac{6}{25}$.

495. (c) Change the percent to a decimal to get 0.80, then multiply: $400 \times 0.80 = 320$.

496. (a) Change the percent to a decimal to get 0.60, then multiply: $390 \times 0.60 = 234$.

SET 32

497. (d) The correct answer is 418.74.

498. (d) Convert the percent to a decimal, so that it becomes 3.0. Now multiply: $20 \times 3.0 = 60$.

499. (a) The correct answer is 19.2.

500. (c) 26% is equal to $\frac{26}{100}$. Changed to a decimal, the value is: $26 \div 100 = 0.26$. Or simply drop the percent sign and move the decimal over two places to the left.

501. (b) Convert the percentage to a decimal: $232 \times 0.14 = 32.48$.

502. (d) A percentage is a portion of 100 where $x\% = \frac{x}{100}$. So the equation is $\frac{x}{100} = \frac{234}{18,000}$. Cross-multiply: $18,000x = 234 \times 100$. Simplify: $x = \frac{23,400}{18,000}$. Thus $x = 1.3$.

503. (d) Since a percentage is a portion of 100 where $x\% = \frac{x}{100}$, the equation is $\frac{x}{100} = \frac{750}{600}$. Crossmultiply: $600x = 750 \times 100$. Simply: $x = \frac{75,000}{600}$ or $x = 125$.

504. (c) To solve the problem, first change the percent to a decimal: $.072 \times 465 = 33.48$, rounded to the nearest tenth is 33.5.

505. (b) $62.5\% = \frac{62.5}{100}$. You should multiply both the numerator and denominator by 10 to move the decimal point, resulting in $\frac{625}{1000}$ and then factor both the numerator and denominator to find out how far you can reduce the fraction. $\frac{625}{1000} = \frac{(5)(5)(5)(5)}{(5)(5)(5)(8)}$. If you cancel the three 5s that are in both the numerator and denominator, you will get $\frac{5}{8}$.

506. (d) Begin by converting $\frac{7}{40}$ into a decimal: $\frac{7}{40} = 0.1750$. Next multiply by 1 in the form of $\frac{100}{100}$ to convert from decimal form to percent form: $(0.1750)(\frac{100}{100}) = \frac{17.50}{100}$ or 17.50%.

507. (c) In your head, you can quickly multiply this figure by 2 to get 36. Then move the decimal point over one space to the left to get 3.60.

508. (a) With this operation you will get the correct amount, which is Rs. 1.80.

1000 MATH PROBLEMS >>> Answers

509. (d) Move the decimal point over two spaces to the left, then multiply: 0.434 × 15 = 6.51.

510. (c) Move the decimal point two spaces to the left, then multiply: 0.002 × 20 = 0.04.

511. (b) Move the decimal point two spaces to the left, then multiply: 0.44 × 5 = 2.2.

512. (a) To change a percent to a decimal, move the decimal point two places to the left, which yields 0.80 or 0.8 (the zero can be dropped as it is on the right side of the decimal point). 0.8 is the same as $\frac{8}{10}$.

SET 33

513. (b) Change the percent to a decimal: 0.35; then, to find the answer, divide: 14 ÷ 0.35 = 40.

514. (b) The root *cent* means *100* (think of the word "century"), so the word *percent* literally means *per 100 parts*. Thus 25% means 25 out of 100, which can also be expressed as a ratio: 25 : 100.

515. (b) The phrase "out of" expresses a ratio—for example: "4:10 doctors" means "four *out of* ten doctors."

516. (b) 0.13 is equal to $\frac{13}{100}$.

517. (b) *Fraction* means top number divided by bottom number. Thus, for example, $\frac{1}{2}$ = 1 ÷ 2 = 0.5 or 50%.

518. (c) The equation to use is $\frac{x}{100} = \frac{12}{50}$. Cross-multiply to get: 50x = (12)(100), or $x = \frac{1200}{50}$. So x = 24 which means 12 is 24% of 50.

519. (c) The formula here is $\frac{12}{100} = \frac{33}{x}$. The solution is 100 × 33 = 12x. 100 × 33 = 3,300. 3,300 ÷ 12 = 275.

520. (b) The fraction $\frac{4}{25}$ means 25 divided into 4, or 0.16. Change the decimal to a percent by moving the decimal point two spaces to the right and adding the percent sign to get 16%.

521. (c) It is easy to mistake 0.8 for 8%, so you must remember: To get a percent, you must move the decimal point *two* spaces to the right. Since there is nothing on the right of 0.8, you must add a zero, then take on the percent sign: 80%.

522. (a) Remember that $\frac{1}{4}$ = 1 ÷ 4 or 0.25; and 100% = 1. Therefore, 1 × 0.25 = 0.25 = 25%.

523. (b) There has been an increase in price of Rs. 3. Rs. 3 ÷ Rs. 50 = 0.06. This is an increase of 0.06, or 6%.

524. (a) It will help to begin by asking the question: Rs. 55 is 15% of what number? First, change the percent to a decimal: 15% = 0.15. Set up the fraction: $\frac{55}{0.15} = 55 \div 0.15 =$ Rs. 366.60. Now subtract: Rs. 366.60 − 55 = Rs. 311.60 spent on CDs.

525. (b) First, change the percent to a decimal: $3\frac{1}{4}\% = 3.25\% = 0.0325$. Now multiply: 30,600 × 0.0325 = Rs. 994.5. Finally, add: Rs. 30,600 + Rs. 994.50 = Rs. 31,594.50 for Yogita's current salary.

526. (d) The basic cable service fee of Rs. 15 is 75 percent of Rs. 20.

527. (a) This is a multiplication problem involving a percent. Because 30% is equivalent to the decimal 0.3, simply multiply the whole number by the decimal: 0.3 × 340 = 102.

528. (b) The problem asks what percent of 250 is 10? Since $x\% = \frac{x}{100}$, the equation is $\frac{x}{100} = \frac{10}{250}$. Cross-multiply: $250x = (10)(100)$. Simplify: $x = \frac{1000}{250}$ or $x = 4$. Thus 4% of the senior class received full scholarships.

SET 34

529. (c) This is a multiplication problem. Change 20% to a decimal and multiply. 13.85 × 0.2 = 2.77.

530. (c) This percent problem involves finding the whole when the percent is given. 280,000 is 150% of last month's attendance. Convert 150% to a decimal. 150% = 1.5. 280,000 = 1.5 × LMA. Next, divide: 280,000 ÷ 1.5 = 186,666.6666... Round up to the nearest whole number: 186,667.

531. (a) This is a two-step problem. First, find the amount of profit. Convert the percent to a decimal and multiply. 70,000 × 0.18 = 12,600. Next, add the result to the original price. 70,000 + 12,600 = 82,600.

532. (d) To find the percent of decrease, first calculate the amount of the decrease. 1.00 − 0.95 = 0.05. Set up the formula to solve for percent. Since $x\% = \frac{x}{100}$, the equation is $\frac{x}{100} = \frac{0.05}{1.00}$. Cross-multiply: $(1.00)(x) = (0.05)(100)$. Simplify: $x = 5$. There is a 5% decrease.

533. (b) To solve this problem, convert the percent to a decimal. $1\frac{1}{2}\% = 0.015$. Then multiply. 48.50 × 0.015 = 0.7275. Round up to 0.73.

534. (d) To find what percent one number is of another, first write out an equation. Since $x\% = \frac{x}{100}$ the equation is: $\frac{4}{1,600} = \frac{x}{100}$. Now cross-multiply: 4 × 100 = 1,600x. Divide both sides by 1,600 to solve for x: x = 0.25.

535. (b) Set up the problem this way: $\frac{x}{90} = \frac{10}{100}$. If you're getting good at percentages, you may just see the answer, but if you don't, cross-multiply: $10 \times 90 = 100x$. Solve: $x = 9$.

536. (a) The equation is $\frac{3}{16} = \frac{x}{100}$. Cross-multiply: $3 \times 100 = 16x$. Solve for x by dividing by 16: $x = 18.75$.

537. (c) To figure 19% of Rs. 423, use the equation $\frac{x}{423} = \frac{19}{100}$. Cross-multiply: $100x = 423 \times 19$. Solving for x gives you 80.37. Subtract that answer from Rs. 423 to get Rs. 342.63.

538. (c) First, determine the percentage of men by subtracting $66\frac{2}{3}\%$ from 100%, which is $33\frac{1}{3}\%$. The formula is then $100 \times 100 = (33\frac{1}{3})x$, or $x = 300$.

539. (a) $7.50 \times 80 = 100x$. $x = 6.00$

540. (b) $644 \times 100 = 795x$. $x = 81$.

541. (d) $10,000 \times 4.5 = 100x$, $x = 450.00$

542. (b) You can't just take 25% off the original price, because the 10% discount after three years of service is taken off the price that has already been reduced by 15%. Figure the problem in two steps: after the 15% discount the price is Rs. 71.83. Ninety percent of that—subtracting 10%—is Rs. 64.65.

543. (b) This is a two-step problem. First, determine what percent of the trees are not oaks by subtracting. 100%—32% = 68%. Change 68% to a decimal and multiply. $0.68 \times 400 = 272$.

544. (b) In this problem, you must find the whole when a percent is given. 3,000 = 15% of Annual Budget. Change to a decimal. $3,000 = 0.15 \times AB$. Solve the equation. $\frac{3,000}{0.15} = AB = 20,000$.

SET 35

545. (d) To find what percent one number is of another, first write out an equation. Since $x\% = \frac{x}{100}$ the equation is: $\frac{x}{100} = \frac{420}{1200}$. Cross-multiply: $1200x = (420)(100)$. Simplify: $x = \frac{42,000}{1200}$. Thus $x = 35$, which means 35% of the videos are comedies.

546. (c) To find the percent change, first determine the original weight. $65.1 - 5.1 = 60$. Recalling that $x\% = \frac{x}{100}$, the equation becomes $\frac{x}{100} = \frac{5.1}{60}$. Cross-multiply: $60x = (5.1)(100)$. Simplify: $x = \frac{510}{60}$. Thus, $x = 8.5$ which means Tony's weight gain was 8.5%.

1000 MATH PROBLEMS >>> Answers

547. (d) In this problem, you must first determine that, because 575 is the 10%-off price, it is also 90% of the original price. So, 575 is 90% of what number? First, change 90% to a decimal: 0.90. Then: $\frac{575}{0.90}$ = 575 ÷ 0.90 = 638.888. Now round up to Rs. 638.89.

548. (a) To solve this problem, change the percent to a decimal and multiply. 0.0525 × 380 = 19.95.

549. (b) To find the whole when a percent is given, first set up an equation. 42 minutes = 70% of Total Homework.

550. (d) To determine the percent, first determine the original length of the rope. 8 + 7 + 5 = 20. Then set up an equation. Knowing that $x\% = \frac{x}{100}$, the equation is: $\frac{x}{100} = \frac{5}{20}$. Cross-multiply: 20x = (5)(100). Simplify: $x = \frac{500}{20}$. Thus x = 25, which means the shortest piece is 25% of the original rope length.

551. (c) To find the whole when the percent is given, first set up an equation. 200 = 40% of the house payment. Convert the percent to a decimal. 200 = 0.4 × HP. Solve. 200 = 0.4 × 500.

552. (c) Container B holds 12% more than container A. Convert the percent to a decimal and set up the equation. B = .12 × 8 + 8 Solve. B = .96 + 8. B = 8.96.

553. (a) Simply set up the equation in the manner in which the problem is written. Since $x\% = \frac{x}{100}$, the equation is $\frac{x}{100} = \frac{.35}{1.40}$. Cross-multiply: 1.40x = (.35)(100). Simplify: $x = \frac{35}{1.40}$. Thus x = 25, which means thirty-five paise is 25% of Rs. 1.40.

554. (d) This is a two-step problem. First, you must determine the number of employees before the new hiring. 30 = 5% of the Original Work Force. Change the percent to a decimal. 30 = 0.05 × OWF. Solve: 30 = 0.05 × 600. Six hundred represents the number of employees before the new hiring. To find the total workforce now, simply add. 600 + 30 = 630.

555. (b) This is a two-step problem involving multiplication and addition. First, determine how many cards were sold on Saturday. 0.05 × 200 = 10. That leaves 190 cards. Then, find out how many cards were sold on Sunday. 0.10 × 190 = 19. Next, add the cards that were sold. 10 + 19 = 29. Finally, subtract from the original number. 200 − 29 = 171.

556. (d) To find the number when percent is given, change the percent to a decimal then set up the equation. 9.50 = 0.45 × money in wallet. Solve: 9.5 = 0.45 × 21.11.

557. (c) This problem requires both multiplication and addition. First, to determine the amount of the raise, change the percent to a decimal and multiply. 0.0475 × 27,400 = 1,301.5. Then, add this amount to the original salary. 1,301.50 + 27,400 = 28,701.50.

1000 MATH PROBLEMS >>> Answers

558. (c) To find the number of days, multiply. $365 \times 0.15 = 54.75$. Round up to 55.

559. (d) First, determine the percent that the station is NOT playing classical music. Subtract from 100%. $100 - 20 = 80$. Eighty percent of the time the station does NOT play classical music. Then change the percent to a decimal and multiply. $0.8 \times 24 = 19.2$.

560. (a) First change 70% to a decimal, which is 0.7. Then multiply. $70 \times 0.7 = 49$.

SET 36

561. (b) To find the percentage of people who said they rode at least 3 times a week, divide 105 by 150: $105 \div 150 = 0.7$, which is 70%. $0.7 \times 100,000 = 70,000$.

562. (d) Division is used to arrive at a decimal, which can then be rounded to the nearest hundredth and converted to a percentage: $113 \div 215 = 0.5255$. 0.5255 rounded to the nearest hundredth is 0.53, or 53%.

563. (b) There are three steps involved in solving this problem. First, convert 4.5% to a decimal: 0.045. Multiply that by Rs. 26,000 to find out how much the salary increases. Finally, add the result (Rs. 1,170) to the original salary of Rs. 26,000 to find out the new salary, Rs. 27,170.

564. (c) First you must subtract the percentage of the installation cost during construction (1.5%) from the percentage of the installation cost after construction (4%). To do this, begin by converting the percentages into decimals: $4\% = 0.04$; $1.5\% = 0.015$. Now subtract: $0.04 - 0.015 = 0.025$. This is the percentage of the total cost which the homeowner will save. Multiply this by the total cost of the home to find the amount: $0.025 \times$ Rs. $150,000 =$ Rs. 3,750.

565. (d) Eighty out of 100 is 80%. Eighty percent of 30,000 is 24,000.

566. (d) Division is used to arrive at a decimal, which can then be rounded to the nearest hundredth and converted to a percentage: $11,350 \div 21,500 = 0.5279$. 0.5279 rounded to the nearest hundredth is 0.53, or 53%.

567. (b) The problem is solved by first converting a fraction to a decimal, then multiplying $2.5 \times 0.39 = 0.975$, which is rounded to 1.

568. (c) First, find 75% of the pump's maximum rate which is $(.75)(100) = 75$ gallons per minute. Next, find the time required using the equation Rate $\times$ Time = Amount pumped. $75T = 1500$. $T = \frac{1500}{75}$. $T = 200$ minutes.

569. (a) $\frac{1}{3} \times 0.60 = 0.20 = 20\%$

570. (a) 20% of 1,800 calories can be calories from fat. 20% of 1,800 equals $(0.2)(1,800)$, or 360 calories from fat are allowed. Since there are 9 calories in each gram of fat, you should divide 360 by 9 to find that 40 grams of fat are allowed.

1000 MATH PROBLEMS >>> Answers

571. (b) Use the formula beginning with the operation in parentheses: $98 - 32 = 66$. After that, multiply 66 by $\frac{5}{9}$, first multiplying 66 by 5 to get 330. 330 divided by 9 = 36.66, which is rounded up to 36.7.

572. (a) At the end of three hours, the organ still has 100% function. After four hours, it has 80% of that 100%, or (0.8) (1), or 0.8. After five hours, it has 80% of the 80% it had at the end of four hours: (0.8) (0.8) equals 64%. After six hours, it has 80% of the 64% it still had after five hours: (0.8) (0.64) equals 0.512.

573. (c) Divide 135 Hindi speaking nurses by 1,125 total numbers of nurses at the hospital to arrive at .12 or 12%.

574. (c) The problem is solved by dividing 204 by 1,700. The answer, 0.12, is then converted to a percentage.

575. (b) The simplest way to solve this problem is to divide 1 by 1,500. which is 0.0006667, and then count off two decimal places to arrive at the percentage 0.06667%. Since the question asks *about what percentage*, the nearest value is 0.067%.

576. (b) 20% of 15 cc = (0.20) (15) = 3. Adding 3 to 15 gives 18 cc.

SET 37

577. (c) The difference between 220 and this person's age = $220 - 30$, or 190. The maximum heart rate is 90% of this, or (0.9) (190) = 171.

578. (a) 72% of 9,125 = (0.72) (9125) = 6,570 males. If 3 out of 5 males were under 25, then 2 out of 5 (or $\frac{2}{5}$) were 25 or older, so $(\frac{2}{5})$ (6570) = 2,628 male patients 25 or older.

579. (d) 30 ppm of the pollutant would have to be removed to bring the 50 ppm down to 20 ppm. 30 ppm represents 60% of 50 ppm.

580. (a) The drug is 50% effective for half (or 50%) of migraine sufferers, so it eliminates (0.50) (0.50) = 0.25 = 25% of all migraines.

581. (a) Add the number of men and women to get the total number of officers: 200. The number of women, 24, is 12% of 200.

582. (d) The 3rd and 4th quarters are 54% and 16% respectively. This adds to 70%.

583. (c) If 60% of the students had flu previously, 40% had not had the disease. 40% of 220 is 88.

584. (d) Four inches is equal to 16 quarter inches, which is equal to (16) (2 feet) = 32 feet. You could also set up the problem as a proportion, so that $\frac{(\frac{1}{4})}{2 \text{ feet}} = \frac{4}{x}$. Cross-multiplying, you would get $(\frac{1}{4})x = 8$. Multiplying through by 4, you would get $x = 32$.

1000 MATH PROBLEMS >>> Answers

585. (a) Each heart's patient takes $\frac{1}{4}$ hour. Each stroke patient takes $3(\frac{1}{4})$ hour, or $\frac{3}{4}$ hour. The doctor has already spent $10(\frac{1}{4})$ plus $3(\frac{3}{4})$, which equals $\frac{10}{4}$ plus $\frac{9}{4}$, or $\frac{19}{4}$, which equals $4\frac{3}{4}$ hours with patients today. Her 6-hour schedule minus $4\frac{3}{4}$ hours leaves $1\frac{1}{4}$ hours left to see patients. Since each stroke patient takes $\frac{3}{4}$ hour, she has time to treat only one more stroke patient in the $1\frac{1}{4}$ hours remaining.

586. (c) You must break the 92,000 into the amounts mentioned in the policy: 92,000 = 20,000 + 40,000 + 32,000. The amount the policy will pay is (0.8)(20,000) + (0.6)(40,000) + (0.4)(32,000). This is 16,000 + 24,000 + 12,800 which equals Rs. 52,800.

587. (b) First, find 50% of 100 gallons per minute which is (.50)(100) = 50 gallons per minute. Next, convert the units of time from minutes to hours. 50 gallons per minute × 60 minutes = 3,000 gallons per hour. Now, use the formula Amount pumped = Rate × Time and plug in the numbers. Amount pumped = (3,000)(6) = 18,000 gallons or $1.80 \times (10^4)$ gallons.

588. (b) The number of papers graded is arrived at by multiplying the rate for each grader by the time spent by each grader. Meera grades 5 papers an hour for 3 hours, or 15 papers; Joe grades 4 papers an hour for 2 hours, or 8 papers, so together they grade 23 papers. Because there are 50 papers, the percentage graded is $\frac{23}{59}$ which is equal to 46%.

589. (c) In one hour Dinesh can do 40% of 40 problems = (.40)(40) = 16 problems. 16 problems is to 40 problems as 1 hour is to x hours: $\frac{16}{40} = \frac{1}{x}$. Cross-multiplying: $16x = 40$. Simplifying: $x = \frac{40}{16} = 2.5$ hrs.

590. (c) First find out how much the population will increase by multiplying 2,500 × 0.03 = 75. Then, add this amount to the current population to get the answer, or 2,500 + 75 = 2,575.

591. (a) If the gas station is 57.8% of the way to the Kapoor family's relative's house, it is (.578)(75) = 43.35 miles from the Kapoor's house. To solve, subtract this distance from the total distance, 75 − 43.35 = 31.65 miles. The gas station is about $31\frac{2}{3}$ miles from the Kapoor family's relative's house.

592. (d) 13% had not read books; therefore, 87% had. 87% is equal to 0.87. 0.87 × 2,500 = 2,175 people.

1000 MATH PROBLEMS >>> Answers

SET 38

593. (d) Change the percent to a decimal: 4% = 0.04. Now multiply: 500 × 0.04 = 20.

594. (a) Multiply the percentages by one another (30% = 0.30; 15% = 0.15.) 0.30 × 0.15 = 0.045 or 4.5%.

595. (d) If 20% are eating the special, 80% are not. This means 40 people represent 80% of the number of people eating at the restaurant. So $.80x = 40$ or $x = \frac{40}{.80}$. Thus $x = 50$ people.

596. (d) An average of 90% is needed of a total of 500 points: 500 × 0.90 = 450, so 450 points are needed. Add all the other test scores together: 95 + 85 + 88 + 84 = 352. Now subtract that total from the total needed, in order to see what score the student must make to reach 90%: 450 − 352 = 98.

597. (c) Adding 7.8 (electrical equipment) and 7.3 (other equipment) is the way to arrive at the correct response of 15.1.

598. (b) Smoking materials account for only 6.7% of the fires but for 28.9% of the deaths.

599. (a) The easiest way to solve this problem is to convert $5\frac{1}{2}\%$ to a decimal: 0.055. The multiply. 200 × 0.055 = 11.

600. (d) Let E = The estimate. One-fifth more than the estimate means $\frac{6}{5}$ or 120% of E, so 600,000 equals (1.20) (E). Dividing both sides by 1.2, we get E equals 500,000.

601. (a) A ratio is given: 3 out of 12. Divide to find the decimal equivalent: 3 divided by 12 = 0.25. Convert to percent form by multiplying by 1 in the form of $\frac{100}{100}$: $(0.25)(\frac{100}{100}) = \frac{25}{100}$ or 25%.

602. (b) This is a multiplication problem involving a percent. 40% is equal to 0.4, so you must multiply the whole number by the decimal: 0.4 × 8 = 3.2.

603. (c) The problem asks, what percent of 34 is 15? To solve the problem, create the fraction $\frac{15}{34}$ and change this to a decimal: $\frac{15}{34} = 0.44$. Now change the decimal to a percent: 44%.

604. (d) First, change 15% to a decimal and multiply: 0.15 × Rs. 26,000 = Rs. 3,900.

605. (c) This percent problem asks that you find the whole when only a percent is known. Convert 325% to a decimal. 325% = 3.25. Now multiply the amount Patty usually spends by the percentage of increase Rs. 4.75 × 3.25 = Rs. 15.44. Since Patty spent Rs. 15.44 more than she usually spends, you must add the amount she usually spends to the amount of increase : Rs. 4.75 + Rs. 15.44 = Rs. 20.19.

1000 MATH PROBLEMS >>> Answers

606. (c) First, find the amount of weight Hemant has lost since 1996: 430 kgs − 337 kgs = 93 kgs. Now create a fraction and solve for the percent: $\frac{93}{430}$ = 93 ÷ 430 = 0.216. Convert the decimal to a percent: 21.6% and round up to 22%.

607. (c) To solve this problem, first change the percent to a decimal. $5\frac{1}{2}\%$ = 5.5% or .055. Now multiply the amount Hanif would usually pay by the percentage of discount: Rs. 75 × 0.055 = 4.12. Finally, subtract: Rs. 75 − 4.12 = Rs. 70.88.

608. (b) Convert 94% to a decimal to get 0.94, then multiply: 250 × 0.94 = 235.

SET 39

609. (c) First, change the percents to decimals: 15% = 0.15 and 27% = 0.27. Now multiply: Rs 875 × 0.15 = Rs. 131.25; Rs. 857 × 0.27 = Rs. 236.25. Add these two amounts: Rs 131.25 + Rs. 236.25 = Rs. 367.5. Now subtract this total from the amount the painting sold for Rs. 875 − Rs. 367.5 = Rs 507.5.

610. (c) The congressman's Rs. 332,000 represents 85% of the original advance (100% − 15%). Designate x as the original advance amount, convert percent to decimal form and set up the equation: $0.85x = 332{,}000$; $x = \frac{332{,}000}{0.85}$, or x = Rs. 390,588.

611. (b) In this two-step problem you must first determine what percent of the dresses are NOT designer dresses by subtracting. 100% − 8% = 92%. Now change 92% to a decimal and multiply. 0.92 × 300 = 276.

612. (b) In this problem, you must find the whole when a percent is given. Rs. 12 = 25% of Amit's weekly budget. Change the percent to a decimal and write out a simple equation: 12 = 0.25 × WB. Now solve the equation by isolating the unknown on one side: $\frac{12}{0.25}$ = WB, or WB = 48. So Amit's weekly budget is Rs. 48.

613. (d) To find what percent one number is of another, first write out an equation. Since $x\% = \frac{x}{100}$ the equation is $\frac{x}{100} = \frac{4}{26}$. Cross-multiply to get: $26x = (4)(100)$, or $x = \frac{400}{26}$, which makes x = 15.4. So approximately 15% of the pies are peach. (Check your answer: 26 × 0.15 = 3.9; round up to 4.)

614. (c) Mona had Rs. 35 prior to this week (Rs. 40 − Rs 5 = Rs 35). So: $\frac{x}{100} = \frac{5}{35}$. By cross-multiplying we get $35x$ (5) (100), $x = \frac{500}{35}$ or x = 14.3. Mona increased her sandwich fund by 14.3% this week. (To check your work, take Rs. 35 × 0.143 = Rs 5.)

615. (d) This is a basic percent problem. Change the percent to a decimal to get 0.15. Now multiply: Rs. 26.50 × 0.15 = Rs. 3.97.

1000 MATH PROBLEMS >>> Answers

616. (c) Because Rs 375 is the 20%-off sale price, it is also 80% of the original price. So, 375 is 80% of what number? First, change the percent to a decimal: 0.80. Then: $\frac{375}{0.80}$ = Rs 375 ÷ 0.80, or Rs. 468.75. (To check your math: 468.75 × 0.80 = Rs. 375.)

617. (c) To solve this problem, first multiply the cost of 1 doll by 6, so Babita buys 6 dolls originally valued at Rs. 1,800. Now change the percent to a decimal (3.5% = 0.035). Rs. 1,800 × 0.035 = Rs. 63, for the amount Babita saves.

618. (d) First, find out how many rooms Manish has already cleaned. Change the percent to a decimal: 40% = 0.4. (Remember zeros to the right of the digit that follows the decimal point do not change the value). Now take 32 × 0.4 = 12.8, which you can round up to 13. Now, subtract: 32 rooms − 13 rooms already cleaned = 19 rooms left to clean. Alternatively, you know that if Manish has cleaned 40% of the 32 rooms, he still has 60% left to go: 32 × 0.6 = 19.2 or about 19 rooms.

619. (c) To determine the percent, first determine the original length of the piece of string cheese: 6 + 4 + 3 = 13 inches. Then set up an equation. Knowing that $x\% = \frac{x}{100}$, the equation is : $\frac{x}{100} = \frac{3}{13}$. Cross multiply to get approximately 23%.

620. (b) To find the whole when the percent is given, first change the percent to a decimal, then multiply: 70% = 0.70. Rs. 450 × 0.70 = Rs. 315.

621. (b) Simply set up the equation in the manner in which the problem is written. Since $x\% = \frac{x}{100}$, the equation is $\frac{x}{100} = \frac{0.40}{1.30}$. Cross-multiply : 1.30$x$ = (.40)(100). Simplify : $x = \frac{40}{1.30}$. Thus x = 30.7, which means that forty paise is about 31% of Rs. 1.30.

622. (a) Change the ratio to a fraction, 3 to 5 = $\frac{3}{5}$. Now change the fraction to a decimal: 3 ÷ 5 = 0.6. Now change the decimal to a percent: 60% of the male dragons will be without female company.

623. (d) It will help to first change the ratio to a fraction, then to a decimal: 2:5 = $\frac{2}{5}$ = 0.4. Now multiply: 365 × 0.4 = 146 rainy days.

624. (b) This is a several-step problem involving multiplication, addition, and subtraction. First, determine how many cookies Sam ate on Saturday morning: 150 × 0.04 = 6. That leaves 144 cookies. Next, determine how many cookies Sam ate on Saturday afternoon: 144 × 0.15 = 21.6. Now, add together all the cookies Sam ate: 6 + 21.6 = 27.6. Finally, subtract the number Sam ate from the original number. 150 − 27.6 = 122.4 cookies, or (rounded) 122.

1000 MATH PROBLEMS >>> Answers

SET 40

625. (c) Hena and Jenny's initial distance apart equals the sum of the distance each travels in 2.5 hours. Hena travels a distance of (2.5)(2.5) = 6.25 miles, while Jenny travels (4)(2.5) = 10 miles. This means that they were 6.25 + 10 = 16.25 miles apart.

626. (b) Suppose total number of apples in the basket = x. No of rotten apples = 135.
∴ 30% of x = 135 or $\frac{30}{100} \times 3 = 135$. $x = \frac{135 \times 100}{30} = 450$.

627. (a) To figure what percentage 1 is of 8, the formula is $\frac{1}{8} = \frac{x}{100}$. Cross-multiply: 100 = 8x. Divide both sides by 8 to get x = 12.5.

628. (d) To figure the total, the formula is $\frac{4}{x} = \frac{20}{100}$. Cross-multiply: $4 \times 100 = 20x$ or 400 = 20x, or x = 20.

629. (c) To figure what percentage 15 is of 60 (minutes in an honour), the formula is $\frac{15}{60} = \frac{x}{100}$. $15 \times 100 = 60x$. Divide both sides by 60 to get x = 25%.

630. (a) $15 \times 100 = 75x$. x = 20.

631. (c) $80 \times 100 = 320x$. x = 25.

632. (b) $200 \times 78 = 100x$. x = 156.

633. (c) First, you must change the percent to a decimal: 85% = 0.85. Now find the amount at which the property is assessed: Rs. 185,000 × 0.85 = Rs. 157,250. Next, divide to find the number of thousands: Rs. 157,250 ÷ 1000 = 157.25. Finally, find the tax: Rs 24.85 × 157.25 = Rs. 3,907.66.

634. (c) Here we find his marks obtained in three subjects;

Marks obtained in English = 50% of 50 = $\frac{50}{100} \times 50 = 25$.

Marks obtained in mathematics = 60% of 70 = 42.

Marks obtained in Hindi = 75% of 80 = 60.

Total marks obtained = 25 + 42 + 60 = 127.

Total of maximum marks = 50 + 70 + 80 = 200

∴ Aggregate percentage = $\frac{127}{200} \times 100 = 63.5\%$

635. (d) The seller's Rs. 103,000 represents only 93% of the sale price (100% − 7%). The broker's commission is NOT 7% of Rs. 103,000, but rather 7% of the whole sale price. The question is: Rs. 103,000 is 93% of what figure? So, let $x = \frac{103,000}{93} = 110,752.68$, rounded to Rs. 110,753.

1000 MATH PROBLEMS >>> Answers

636. (c) First, change the percents to decimals. Next, find the total commission: Rs. 115,000 × 0.06 = Rs. 6,900. Finally, find salesperson Sunil's cut, which is 55% of the total price (45% having gone to broker Bobby): Rs. 6,900 × 0.55 = Rs. 3,795.

637. (a) First, find $\frac{1}{12}$ of the tax bill: Rs. 18,000 ÷ 12 = 1,500. Now find 3% of the gross receipts: Rs. 75,000 × 0.03 = 2250. Now add: Rs. 1,000 (rent) + Rs. 1,500 (percent of annual tax bill) + Rs. 2,250 (percent of gross receipts) = Rs. 4,750.

638. (b) First find the amount of assessment Rs. 325,000 × 0.90 = Rs. 292.50. Now find the number of thousands : = Rs. 292,500 ÷ 1000 = 292.5. Next, find the tax rate for a whole year: 292.5 × Rs. 2.90 = 848.25. Now find the tax rate for half of the year : Rs. 848.25 ÷ 2 = Rs. 424.13.

639. (a) Production in 1983-84 = 151.8 million tonnes. Production 1978-79 = 132 Million tonnes. Increase in production = (151.8 − 132) = 19.8 million tonnes. ∴ Increase percentage = $\frac{19.8}{132} \times 100$ = 15%.

640. (a) Suppose original consumption of petrol = 100 liters and original price of 100 liters of petrol = Rs. 100. Increased price of 100 litres of petrol = Rs. 100 + Rs. 25. = Rs. 125. ∴ Now Satish spends Rs. 125 to buy 100 liters. Satish spends Rs. 100 to buy $\frac{100 \times 100}{125}$ or .80 liters. ∴ reduction in consumption = (100 − 80)% = 20%.

SECTION 5 : ALGEBRA

SET 41

641. (d) Both sides of an equation are always equal. Therefore, the best synonym is balance scale.

642. (c) A percentage is a portion of 100, or $\frac{x}{100}$. The equation here is $\frac{x}{100} = \frac{12}{50}$, or $12 \times 100 = 50x$. $12 \times 100 = 1,200$; $1,200 \div 50 = 24$; therefore, $x = 24\%$.

643. (a) The problem is solved by first determining that $8n = 40$, then dividing 40 by 8 to get the answer, which is 5.

644. (c) If the number is represented by n, its double is $2n$. Therefore, $n + 2n = 69$; $3n = 69$; $n = 23$.

645. (d) Solve this problem with the following equation: $4x - 12 = 20$; therefore, $4x = 32$, so $x = 8$.

646. (d) Let x = the unknown number. We have $x + 0.50x = 27$ or $1.50x = 27$. Simplify: $x = \frac{27}{1.5}$. Therefore $x = 18$.

647. (d) Let x = the number sought. One-sixth of a certain number becomes $(\frac{1}{6})x$, four more than one-twelfth the number becomes $(4 + (\frac{1}{12})x.)$. Combining terms: $(\frac{1}{6})x = 4 + (\frac{1}{12})x$. Rearranging: $(\frac{1}{6})x - (\frac{1}{12})x = 4$ or $(\frac{1}{6} - \frac{1}{12})x = 4$ which reduces to $(\frac{1}{12})x = 4$. Thus $x = 48$.

648. (a) Let x = the number sought. We have: $x = (\frac{1}{9})(45) - 6$ or $x = -1$.

649. (b) Let x = the number sought. Twelve times one-half of a number becomes $(12)(\frac{1}{2})x$. Thus we have: $(12)(\frac{1}{2})x = 36$ or $6x = 36$ and $x = 6$.

650. (d) Let x = the number sought. To solve this problem break it up into smaller pieces: 21 times 4 = $(21)(4)$, one twelfth of a number = $(\frac{1}{12})x$. Combining terms: $(21)(4)(\frac{1}{12})x = 7$. Simplifying: $(\frac{84}{12})x = 7$, which becomes $7x = 7$, or $x = 1$.

651. (c) Let x = the number sought. Eight and three both taken away from a number means $x - (7 + 3)$, and we have $x - (7 + 3) = 31$ or $x = 31 + 10$. Thus $x = 41$.

652. (d) Let x = the unknown percentage. Restate the problem as $35 = 90x$ or $x = \frac{35}{90}$. Thus $x = 0.38 = 38\%$.

653. (d) Let x = the number sought. Solve this by breaking up the problem into parts: Six less than three times a number = $3x - 6$, four more than twice the number = $2x + 4$. Combining terms: $3x - 6 = 2x + 4$. Simplifying: $3x - 2x = 6 + 4$, or $x = 10$.

654. (d) Let x = the number sought. The word "product" tells us to multiply 16 by one-half x, or $(16)(.5x)$, which we set equal to 136. Therefore, $(16)(.5x)$ 136, which reduces to $8x = 136$, resulting in $x = 17$.

655. (c) Let x = the number sought. Begin by breaking up the problem into smaller pieces: 12 more = 12 +, 30 percent of a number = $0.30x$, and one-half the number = $0.50x$. Next, combine the terms, $12 + 0.30x = 0.50x$, and simplify, $0.20x = 12$. Thus $x = 60$.

656. (b) Let x = the number sought. Begin by breaking up the problem into smaller pieces: two times a number = $2x$, seven times a number = $7x$, seven times a number is taken away from 99 = $99 - 7x$. Next, combine the terms $2x = 99 - 7x$, and simplify, $9x = 99$. Thus $x = 11$.

SET 42

657. (d) Let x = the number sought. Forty-five together with is 45 +, nine-fifths of a number is $\frac{9}{5}x$, twice the number is $2x$. Combining terms: $45 + \frac{9}{5}x = 2x$. Simplifying: $45 = 2x - \frac{9}{5}x$ which becomes $45 = 0.2x$. Thus $x = 225$.

658. (c) The formula for percentages is $\frac{12}{100} = \frac{33}{x}$. The solution is $100 \times 33 = 12x$. $100 \times 33 = 3{,}300$, and $3{,}300 \div 12 = 275$; therefore, $x = 275$.

659. (b) Let x = the number sought. Sixty percent of 770 becomes $0.60(770)$, "seven times what number" we restate as $7x$. Combining terms: $0.60(770) \div 6 = 7x$. Simplifying: $77 = 7x$ or $x = 11$.

660. (b) Let x = the number sought. 19 more than a certain number is 63 means: $x + 19 = 63$ or $x = 63 - 19$. Thus $x = 44$.

661. (d) Let x = the number sought. Begin the solution process by breaking up the problem into smaller parts: a number 3 times larger = $3x$, 10 added to the number = $x + 10$. Combining terms we have: $3x = x + 10$. Simplifying: $2x = 10$, or $x = 5$.

662. (b) Let x = number sought. Eleven and forty-one together means $(11 + 41)$, thus we have: $(11 + 41) \div x = 13$. Cross-multiplying: $13x = 52$ or $x = 4$.

663. (a) $4n + 20 = 72$; $4n = 52$; $n = 13$.

1000 MATH PROBLEMS >>> Answers

664. (d) The equation is simply $50 + 3x = 74$. $3x = 24$; $x = 8$.

665. (a) Let x = the number sought. Four more than three times a number means: $(3x + 4)$. So the expression becomes: $(2)(3x + 4) = 20$. Simplifying: $6x + 8 = 20$ or $6x = 12$. Thus $x = 2$.

666. (c) One of the most vital steps in solving for an unknown in any algebra problem is to isolate the unknown on one side of the equation.

667. (d) Substitute 3 for x in the expression 5 plus $4x$ to determine that y equals 17.

668. (d) Let x = the number sought. The word "product" tells us to multiply 16 by one-half x, or $(16)(.5x)$, which we set equal to 136. Therefore, $(16)(.5x) = 136$, which reduces to $8x = 136$, resulting in $x = 17$.

669. (c) Let x = the number sought. Begin by breaking up the problem into smaller pieces: 12 more = 12+, 30 percent of a number = $.30x$, and one-half the number = $0.50x$. Next, combine the terms, $12 + .30x = 0.50x$, and simplify, $0.20x = 12$. Thus $x = 60$.

670. (b) Let x = the number sought. Begin by breaking up the problem into smaller pieces: two times a number = $2x$, seven times a number = $7x$, seven times a number is taken away from 99 = $99 - 7x$. Next, combine the terms, $2x = 99 - 7x$, and simplify, $9x = 99$. Thus $x = 11$.

671. (a) Since the solution to the problem $x + 25 = 13$ is –12, choices b, c, and d are all too large to be correct.

672. (d) The first step in solving this problem is to add the fractions to get the sum of $\frac{4x}{4}$. This fraction reduces to x.

SET 43

673. (b) Because the integers must be even, the equation $n + (n + 2) + (n + 4) = 30$ is used. This gives $3n + 6 = 30$; $3n = 24$; $n = 8$. Therefore, 8 is the first number in the series. Option a, (9, 10, 11) would work, but the numbers aren't even integers.

674. (c) The solution is as follows: $[6(3)] - [6(-2)] \div 9$. The equation then becomes $18 - (-12) \div 9$, and then, because two minuses become a plus, $30 \div 9 = 3\frac{1}{3}$.

675. (a) Cross-multiplying: $(2x)(48) = (16)(12)$; $96x = 192$. Thus $x = 2$.

676. (a) If the inequality is solved as an equation, the largest value that fulfills the inequality is found. Therefore, $3x - 14 = 3$; $3x = 17$; $x = 5\frac{2}{3}$. Any number smaller than this will fulfill the inequality. The only number less than $5\frac{2}{3}$ is 4.

677. (d) $(x^2 + 4x + 4)$ factors into $(x + 2)(x + 2)$. Therefore, one of the $(x + 2)$ terms can be canceled with the denominator. This leaves $(x + 2)$.

1000 MATH PROBLEMS ▸▸▸ Answers

678. (a) $x(x + 3) = x(x) + (3x) = x^2 + 3x$.

679. (b) Let $x =$ the number sought. The statement twenty-three added to a number means $x + 23$, ninety-nine is the result is the same as saying $= 99$. Therefore, we have: $x + 23 = 99$ or $x = 99 - 23$. Thus $x = 76$.

680. (a) $2x(3xy + y) = 2x(3xy) + 2x(y) = 6x^2y + 2xy$.

681. (b) $x^2 - 4x + 4$ is equal to $(x - 2)^2$. When this is divided by $x - 2$, it simplifies to $x - 2$.

682. (b) To square y, multiply $y \times y$.

683. (d) x times x^2 is x^3; x times y is xy, so the solution to the problem is $3x^3 + xy$.

684. (a) When the first term is expanded, the exponents are multiplied. This gives $8x^9y^3$. When this is multiplied by the second term, the exponents are added, giving $32x^{11}y^5$.

685. (c) This equation is a proportion, expressing the equivalence of fractions.

686. (c) The quotient of two numbers is $x \div y$. When a third number, z, is added, the result is: $x \div y + z$.

687. (d) To solve this problem, you must first find the common denominator, which is 6. The equation then becomes $\frac{3x}{6} + \frac{x}{6} = 4$; then, $\frac{4x}{6} = 4$; and then $4x = 24$, $x = 6$.

688. (b) Raise the fraction $\frac{2}{9}$ to 54ths by multiplying both numerator and denominator by 6. $\frac{12}{54} = \frac{2}{9}$.

SET 44

689. (b) $\frac{1}{3}x + 3 = 8$. In order to solve the equation, all numbers need to be on one side and all x values on the other. Therefore, $\frac{1}{3}x = 5$; $x = 15$.

690. (c) Seven is added to both sides of the equation, leaving $2x = 11$. Eleven is divided by 2 to give $\frac{11}{2}$.

691. (b) Eight is substituted for x. $8^2 = 8 \times 8 = 64$. $\frac{64}{4} = 16$; $16 - 2 = 14$.

692. (a) $x = \frac{1}{16} \times 54$, which is equivalent to $54 \div 16$, which is 3.375.

693. (c) Seven is added to both sides of the equation, giving $1.5x = 19.5$. $19.5 \div 1.5 = 13$.

694. (d) All are possible solutions to the inequality; however, the largest solution will occur when $\frac{1}{3}x - 3 = 5$. Therefore, $\frac{1}{3}x = 8$; $x = 24$.

695. (b) The equation is $y = mx + b$, where m is the slope of the line and b is the y intercept.

1000 MATH PROBLEMS >>> Answers

696. (b) The simplest way to solve this problem is to cancel the *a* term that occurs in both the numerator and denominator. This leaves $\frac{(b-c)}{bc}$. This is $\frac{(4-(-2))}{4(-2)}$, which simplifies to $-\frac{3}{4}$.

697. (d) After finding a common denominator, the equation becomes $\frac{4}{12}x + \frac{3}{12}x = 3$. By cross-multiplying, the common denominator can be found. This gives $\frac{7x}{12} = 3$; $x = \frac{36}{7} = 5\frac{1}{7}$.

698. (c) Slope is equal to the change in *y* values ÷ the change in *x* values. Therefore, $\frac{(3-(-1))}{(2-0)} = \frac{4}{2} = 2$. The intercept is found by putting 0 in for *x* in the equation $y = 2x + b$. $-1 = 2(0) + b$; $b = -1$. Therefore, the equation is $y = 2x - 1$.

699. (a) Let x = the number sought. Working in reverse order we have: the sum of that number and six becomes $(x + 6)$, the product of three and a number becomes $3x$. Combining terms: $(x + 6) - 3x = 0$. Simplifying: $2x = 6$ or $x = 3$.

700. (b) $\frac{1}{3}x + 3 = 8$. In order to solve the equation, all numbers need to be on one side and all *x* values on the other. Therefore, $\frac{1}{3}x = 5$; $x = 15$.

701. (b) Two equations are used. T = 4O, and T + O = 10. This gives 5O = 10, and O = 2. Therefore, T = 8. The number is 82.

702. (a) The sum of three numbers means $(a + b + c)$, the sum of their reciprocals means $(\frac{1}{a}+\frac{1}{b}+\frac{1}{c})$. Combining terms: $(a + b + c)(\frac{1}{a}+\frac{1}{b}+\frac{1}{c})$. Thus choice **a** is the correct answer.

703. (d) Add like terms: $6x + 3x = 9x$. $5y - 5y = 0$.

704. (b) Let x = the number sought. Begin by breaking the problem into parts: *88 is the result* becomes = 88. *One-half of the sum of 24 and a number* becomes $0.5(24 + x)$. *Three times a number* becomes $3x$. Combining terms: $3x - 0.5(24 + x) = 88$. Simplifying: $3x - 12 - 0.5x = 88$ which reduces to $2.5x = 100$. Thus $x = 40$.

SET 45

705. (b) The variable is a symbol that stands for any number under discussion.

706. (d) The total length of pole is = $\frac{1}{3}x + \frac{1}{4}x + 5 = x$. Multiplying both sides by 12. (The l.c.m. of 4, 3) that is $4x + 3x + 60 = 12x$, $5x = 60$, $x = 12$.

707. (b) We want to know D = difference in total revenue. Let A = total revenue at Rs. 6 each, and B = total revenue at Rs. 8 each. Therefore: D = A − B. We are given A = 6 × 400 = 2,400 and B = 8 × 250 = 2,000. Substituting: D = 2,400 − 2,000. Thus D = 400 Rupees.

708. (c) We are asked to find B = amount of tips Bina is to receive. We know from the information given that: T + R + B = 48, T = 3R, and B = 4R. The latter can be rearranged into the more useful: R = B ÷ 4. Substituting: 3R + R + B = 48. This becomes: 3 (B ÷ 4) + (B ÷ 4) + B = 48. Simplifying: (3 + 1) (B ÷ 4) + B = 48 which further simplifies to: 4(B ÷ 4) + B = 48, or 2B = 48. Thus B = Rs. 24.

709. (c) F is replaced by 95 in the equation. Thirty-two is subtracted from both sides, leaving $63 = \frac{9}{5}C$. $C = 63 ÷ \frac{9}{5} = 35$.

710. (c) The total bill is the sum of the three people's meals. We are solving for the father's meal. The mother's meal is $\frac{5}{4}F$, and the child's meal is $\frac{3}{4}F$. Therefore, $F + \frac{5}{4}F + \frac{3}{4}F = 240$. This simplifies to 3F = 480 F = Rs. 80.

711. (b) We are solving for x = the cost per kg of the final mixture. We know that the total cost of the final mixture equals the total cost of the dried fruit plus the total cost of the nuts, of C = F + N. In terms of x: C is the cost per kg times the number of kgs in the mixture, or C = 7.5x. Substituting: 7.5x = 300(6) + 700(1.5). Simplifying: 7.5x = 1800 + 1050 or $x = \frac{2850}{7.5}$ = Rs. 380/-

712. (d) We want to know R = helicopter's speed in mph. To solve this problem recall that: Rate × Time = Distance. It is given that T = 6:17 − 6:02 = 15 minutes = 0.25 hour and D = 20 miles. Substituting: R × 0.25 = 20. Simplifying: R = 20 ÷ 0.25. Thus R = 80 mph.

713. (c) 12 × 5% + 4 × 4% = x times 16 ; x = 4.75%.

714. (b) Let Dinesh's rate = x. Dinesh's rate multiplied by his travel time equals the distance he travels; this equals Arun's rate multiplied by his travel time; 2x = D = 3(x − 5). Therefore, 2x = 3x − 15 or x = 15 mph.

715. (d) The equation to describe this situation is $\frac{10 \text{ fish}}{\text{hour}}$ (2 hours) = $\frac{2.5 \text{ fish}}{\text{hour}} t$; 20 = 2.5$t$; t = 8 hours.

716. (b) Solve this problem by finding the weight of each portion. The sum of the weights of the initial corn is equal to the weight of the final mixture. Therefore, (20 bushels) $\frac{56 \text{ kgs}}{\text{bushel}}$ + (x bushels) $\frac{50 \text{ kgs}}{\text{bushel}}$ = ((20 + x) bushels) $\frac{54 \text{ kgs}}{\text{bushel}}$. Thus 20 × 56 + 50$x$ = (x + 20) × 54.

717. (d) The total sales equal the sum of Lina and Hira's sales or: L + J = 36. Since Lina sold three less than twice Hira's total, L = 2J − 3. The equation (2H − 3) + H = 36 models this situation. This gives 3H = 39; H = 13.

1000 MATH PROBLEMS >>> Answers

718. (d) The amount done in one day must be found for each person. Giri strings $\frac{1}{4}$ of a fence in a day, and Mini strings $\frac{1}{3}$ of a fence in a day : $\frac{1}{4} + \frac{1}{3} = \frac{1}{x}$; multiplying both sides by $12x$ yields $3x + 4x = 12$; $x = \frac{12}{7} = 1\frac{5}{7}$ days.

719. (b) $1\frac{1}{2}$ cups equals $\frac{3}{2}$ cups. The ratio is 6 people to 4 people, which is equal to the ratio of $\frac{3}{2}$. By cross-multiplying, we get $6(\frac{3}{2})$ equals $4x$, or 9 equals $4x$. Dividing both sides by 9, we get $\frac{9}{4}$, or $2\frac{1}{4}$ cups.

720. (c) One gallon of 8% solution plus x amount of water is equal to $(1 + x)$ amount of 2% solution. Since pure water is 0% salt, we have: $(1)(0.08) + x(0.00) = (1 + x)(0.02)$ and the equation simplifies to $0.08 = 0.02 + 0.02x$; $0.02x = 0.06$; $x = \frac{0.06}{0.02}$. Thus $x = 3$ gallons.

SET 46

721. (b) The equation $2x + x + \frac{1}{2}x = 20$ models this situation. Therefore, $3\frac{1}{2}x = 20$; $x = 5\frac{5}{7}$ feet.

722. (c) Raj is four times as old as Ram means $R_1 = R_2$, Ram is one-third as old as Ravi means $R_2 = \frac{1}{3}R_3$. We are given $R_3 = 18$. Working backwards we have: $R_2 = \frac{1}{3}(18) = 6$; $R_2 = 4(6) = 24$. The sum of their ages $= R + R + R = 24 + 6 + 18 = 48$.

723. (b) The algebraic equations used are $\frac{1}{4}(K - 5) = L - 5$ and $L + K = 110$. $K = 110 - L$; this is put in the first equation to get $\frac{1}{4}(105 - L) = L - 5$. Solve for L; $L = 25$.

724. (b) To solve this problem set up the proportion 3 is to 25 as x is to 40: $\frac{3}{25} = \frac{x}{40}$. Cross-multiplying: $(3)(40) = 25(x)$. Solving for x gives 4.8, but since coolers must be whole numbers, this number is rounded up to 5.

725. (b) A ratio is set up: $\frac{1.5}{14} = \frac{2.25}{x}$. Solving for x gives 21 laddoos.

726. (a) Two equations are used. $A + B + C = 25$, and $A = C = 2B$. This gives $5B = 25$, and $B = 5$.

727. (c) The problem is to find x = number of gallons of 75% antifreeze. We know the final mixture is equal to the sum of the two solutions or $M = A + B$. In terms

213

of x, M = 0.50 $(x + 4)$, A = 0.75x, and B = 0.30(4). Substituting, 0.50$(x + 4)$ = 0.75x + 1.20. Simplifying, 0.50x + 2 = 0.75x + 1.2 which reduces to 0.25x = 0.80. Thus x = 3.2 gallons.

728. (d) We want to find x = the number of kgs of concrete with 14% cement. We know the final concrete mixture equals the sum of the two solutions or M = A + B. In terms of x, M = .11 $(x + 150)$, A = 0.14x, and B = .06(150). Substituting, 0.11 $(x + 150)$ = 0.14x + 9. Simplifying, 0.11x + 16.5 = 0.14x + 9 which reduces to .03x = 7.5. Thus x = 250 kgs.

729. (a) The problem is to find x = number of kgs of Apples. We know that total fruit purchased equals the sum of bananas and apples, or F = B + A. In terms of x, F = 2$(x + 7)$, B = 7(0.50), and A = 4x. Substituting, 2 $(x + 7)$ = 3.50 + 4x. Simplifying, 2x + 14 = 3.5 + 4x, which reduces to 2x = 105. Thus x = 5.25 kgs.

730. (b) The problem is to find x = number of rental months needed to make the costs equal. This occurs when purchase cost equals rental cost, or P = R. We are given P = 40,000 and R = 500 + 5000 × 2500x. Substituting, 40000 = 50000 + 2500x which reduces to 2500x = 3500. Thus x = 14 months.

731. (b) Two equations are used. E = M + 10, and 20E + 25M = 1460. This gives 20M + 200 + 25M = 1460; 45M = 1260; M = 28.

732. (b) We are asked to find x = total fruit cost in rupees. This is the sum of the cost of the apples and the cost of the grapes or x = cost of apples + cost of grapes. Each fruit's cost is found by multiplying the price per kg by the number of kgs so the cost of apples = (0.90)(47), and the cost of grapes = (1.49) (19). Therefore x = (0.90)(47) + (1.49)(19) which reduces to x = 42.30 + 28.31 or x = Rs. 70.61.

733. (c) The unknown I = the amount of interest earned. To solve this problem it is necessary to use the formula given: Interest = Principal × Rate × Time (I = PRT). First, change the percent to a decimal: $7\frac{1}{8}$ = 0.7125. Next note that nine months = $\frac{9}{12}$ or $\frac{3}{4}$ or 75% (0.75) of a year. Substituting, we have I = (Rs. 767)(0.07125)(0.75). Thus I = Rs. 40.99.

734. (a) We are asked to find F = Vicky's age. Begin the solution by breaking the problem into parts: Vicky is half the age of Mony becomes V = $\frac{1}{2}$M, Mony is one-third as old as Sanju becomes M = $\frac{1}{3}$S, and Sanjay is half the neighbour's age becomes S = $\frac{1}{2}$N. We know the neighbour's age is 24 or N = 24. Substituting and working backwards through the problem: S = $\frac{1}{2}$(24) = 12, M = $\frac{1}{3}$(12) = 4, V = $\frac{1}{2}$(4). Thus V = 2, Vicky is two years old.

735. (b) Let x = the unknown quantity of each denomination. We know that all the coins total Rs. 8.20 and each denomination is multiplied by the same number,

1000 MATH PROBLEMS >>> Answers

x. Therefore $0.25x + 0.10x + 0.05x + 0.01x = 8.20$. This reduces to $(0.25 + 0.10 + 0.05 + 0.01)x = 8.20$, or $0.41x = 8.20$. Thus $x = 20$ coins in each denomination.

736. (a) The problem is to find $x =$ number of kgs of nuts costings Rs 7 per kg. The total cost of the mixture equals the sum of the cost for each type of nut or: $M = A + B$. Where $A = 7x$, $B = 3(6)$, and $M = 4(6 + x)$. Substituting: $4(6 + x) = 7x + 18$. Simplifying: $24 + 4x = 7x + 18$, which becomes $24 - 18 = 7x - 4x$; or $3x = 6$. Thus $x = 2$ kgs.

SET 47

737. (b) Let $x =$ The amount of interest Savita earns in one year. Substituting: $I = (385)(.0485)(1)$. Thus $I =$ Rs. 18.67.

738. (a) Let R = Veena's average speed. Recall that for uniform motion distance = Rate × Time or $D = RT$. Substituting: $220 = R(5)$ or $R = \frac{220}{5}$. Thus $R = 44$ mph.

739. (c) The problem is probably going to ask you to solve for the unknown, x, which stands for Arvind's age.

740. (c) The ratio is $\frac{(12 \text{ cc})}{(100 \text{ kgs})} = \frac{(x \text{ cc})}{175 \text{ kgs}}$, where x is the number of cc's per 175 kgs. You must multiply both sides by 175 to get $(175)(\frac{12}{100}) = x$, so x equals 21.

741. (a) Set up an equation, first of all, to find out how many *months* it will take for the lease rate plus deposit to equal the purchase price: $340x + 1,000 = 31,600$. Solving for x gives you 90, but remember that this is 90 *months*, while the problems asks for *years*. Divide 90 by 12 to get 7.5 years.

742. (a) The problem is to find $x =$ number of kgs of $\frac{65}{35}$ solder. We arbitrarily choose to solve for Tin, but the same method can be used to solve for Lead. We know the final mixture equals the sum of the two types of solder, or $M = A + B$. In terms of x, $M = 0.50(80)$, $A = 0.65x$, and $B = 0.25(80 - x)$. Substituting, $40 = 0.65x + 0.25(80 - x)$. Simplifying, $40 = 0.65x + 20 - 0.25x$, which reduces to $0.40x = 20$. Thus $x = 50$ kgs.

743. (b) $150x = (100)(1)$, where x is the part of a mile a jogger has to go to burn the calories a walker burns in 1 mile. If you divide both sides of this equation by 150, you get $x = \frac{100}{150} = \frac{[2(50)]}{[3(50)]}$. Cancelling the 50s, you get $\frac{2}{3}$. This means that a jogger has to jog only $\frac{2}{3}$ of a mile to burn the same number of calories a walker burns in a mile of brisk walking.

744. (b) 5% of 1 litre $= (0.05)(1) = (0.02)x$, where x is the total amount of water in the resulting 2% solution. Solving for x, you get 2.5. Subtracting the 1 litre of water already present in the 5% solution, you will find that 1.5 (= 2.5 − 1) litres need to be added.

1000 MATH PROBLEMS >>> Answers

745. (c) The problem is to find A = Akshay's present age in years. Begin by breaking the problem up into smaller parts: Akshay will be twice Sunil's age in 3 years becomes A + 3 = 2S; Sunil will be 40 becomes S = 40. Substituting: A + 3 = 2(40). Simplifying: A = 80 − 3, or A = 77 years old.

746. (c) Let E = emergency room cost; H = hospice cost = $(\frac{1}{4})$E; N = home nursing cost = 2H = 2$(\frac{1}{4})$E = $(\frac{1}{2})$E. The total bill = E + H + N = E + $(\frac{1}{4})$E + $(\frac{2}{4})$E = 140,000. So $(\frac{4}{4})$E + $(\frac{1}{4})$E + $(\frac{2}{4})$E = 140,000, so $(\frac{7}{4})$E = 140,000. Multiplying both sides by $(\frac{4}{7})$ to solve for E, we get E = 140,000$(\frac{4}{7})$ = 80,000. Therefore H = $(\frac{1}{4})$ E = $(\frac{1}{4})$ 80,000 = 20,000 and N = 2H = 2(20,000) = 40,000.

747. (d) 3W equals water coming in, W equals water going out. 3W minus W equals 11,400, which implies that W is equal to 5,700 and 3W is equal to 17,100.

748. (d) Let T equal Tony age; S equal Sam's age, which is 3T; R equal Ram's age, which is $\frac{S}{2}$, or $\frac{3T}{2}$. The sum of the ages is 55, which is $\frac{3T}{2}$ plus 3T plus T, which is equal to $\frac{3T}{2}$ plus $\frac{6T}{2}$ plus $\frac{2T}{2}$, which is equal to $\frac{(3T+6T+2T)}{2}$, or $\frac{11T}{2}$. Now multiply both sides of 55 equals $\frac{11T}{2}$ by 2 to get 110 equals 11T. Divide through by 11 to get 10 equals T. That is Tony's age, so Sam is 3T, or 3(10), or 30 years old, and Ram is $\frac{3T}{2}$, which is $\frac{3(10)}{2}$, or $\frac{30}{2}$, or 15 years old.

749. (d) We are asked to find x = minutes until the boat disappears. Recall that Distance = Rate × Time or D = RT or T = $\frac{D}{R}$. We are given D = 0.5 mile and R = 20 mph. Substituting: T = $\frac{0.5}{20}$ = 0.025 hour. Convert the units of time by establishing the ratio. .025 hours is to 1 hour as x minutes is to 60 minutes or $\frac{.025}{1} = \frac{x}{60}$. Cross-multiplying: (.025)(60) = (1)(x) which simplified to x = 1.5 minutes.

750. (a) First you find out how long the entire hike can be, based on the rate at which the hikers are using their supplies. $\frac{\frac{2}{5}}{3} = \frac{1}{x}$, where 1 is the total amount of supplies and x is the number of days for the whole hike. Cross-multiplying, you get $\frac{2x}{5}$ = 3, so that $x = \frac{(3)(5)}{2}$, or $7\frac{1}{2}$ days for the length of the entire hike. This means that the hikers could go forward for 3.75 days altogether before they would have to turn around. They have already hiked for 3 days. 3.75 minus 3 equals 0.75 for the amount of time they can now go forward before having to turn around.

1000 MATH PROBLEMS >>> Answers

751. (b) An algebraic equation should be used: $K - 20 = \frac{1}{2}(M - 20)$; $K = 40$. There, $M = 60$.

752. (a) You know the ratio of Khanna's charge to Jean's charge is 3 to 4, or $\frac{3}{4}$. To find what Shah charges, you must use the equation $\frac{3}{4} = \frac{36}{x}$, or $3x = 4(36)$. $4(36) = 144$, which is then divided by 3 to arrive at $x = 48$.

SET 48

753. (b) This uses two algebraic equations to solve for the age. Anju (A) and her grandfather (G) have a sum of ages of 110 years. Therefore, $A + G = 110$. Anju was $\frac{1}{3}$ as young as her grandfather 15 years ago. Therefore, $A - 15 = \frac{1}{3}(G - 15)$. Either equation can be solved for A or G and substituted into the other. $A = 110 - G$; $110 - G - 15 = \frac{1}{3}G - 5$; $100 = \frac{4}{3}G$; $G = 75$.

754. (c) Let x = the number of oranges left in the basket. Three more than seven times as many oranges as five is: $7(5) + 3 = 38$. Removing five leaves $x = 38 - 5 = 33$ oranges.

755. (b) The problem is to find x = number of ounces of candy costing Re. 1 (or 100 paise) per ounce. The total cost of the mixture equals the sum of the cost for each type of candy or: $M = A + B$, where $A = 100x$, $B = 70(6)$, and $M = 80(6 + x)$. Substituting: $80(6 + x) = 100x + 420$. Simplifying: $480 + 80x = 100x + 420$, which becomes $480 - 420 = 100x - 80x$, or $20x = 60$. Thus, $x = 3$ ounces.

756. (a) Let x = the number of hours to wash and wax the car if both work together. In 1 hour Dinoo can do $\frac{1}{4}$ of the job, while Giri can do $\frac{1}{3}$ of the job. In terms of x this means: $\frac{1}{4}x + \frac{1}{3}x = 1$ (where 1 represents 100% of the job). Simplifying: $(\frac{1}{4} + \frac{1}{3})x = 1$ or $(\frac{3}{12} + \frac{4}{12})x = 1$. Thus $\frac{7}{12}x = 1$ or $x = \frac{12}{7} = 1.7$ hours.

757. (b) Let T = The time it takes Shaila to walk to the store. Since she walks at a uniform rate we can use the formula Distance = Rate × Time or $D = RT$. Substituting: $5 = 3T$ or $T = \frac{5}{3}$. Thus $T = 1.67$ hours.

758. (a) Let D = the time Dinesh arrived before class. Choosing to represent time before class as a negative number we have: Sam arrived 10 minutes early means $S = -10$, Dinesh came in 4 minutes after Mona means $D = M + 4$, Mona, who was half as early as Sam means $M = \frac{1}{2}S$. Substituting: $M = -5$, so $D = -5 + 4 = -1$. Thus $D = 1$ minute before class time.

759. (d) Substituting: $5 = (45)R(\frac{1}{12})$ or $5 = (45)R(.0833)$. Thus $R = 1.33 = 133\%$.

217

1000 MATH PROBLEMS >>> Answers

760. (b) Let x = the number of people remaining in the room. We have: $x = 12 - (\frac{2}{3}(12) + 3)$ or $x = 12 - (8 + 3) = 12 - 11$. Thus $x = 1$ person.

761. (b) Let x = The height of the ceiling. After converting 24 inches = 2 feet, we have $2 = 0.20x$ or $x = 10$ feet.

762. (d) Let x = The number of hours it takes Mamta to complete the job. In 1 hour the neighbour can do $\frac{1}{38}$ of the job, while Mamta can do $\frac{1}{x}$ of the job. Working together it takes 22 hours to complete 100% of the job or: $\frac{1}{38}(22) + \frac{1}{x}(22) = 1$ (where 1 represents 100% of the job). Simplifying: $\frac{22}{38} + \frac{22}{x} = 1$ or $\frac{22}{x} = 1 - \frac{22}{38}$, which reduces to $\frac{22}{x} = \frac{16}{38}$. Cross-multiplying: $16x = (22)(38)$ or $x = 52.25$ hours.

763. (c) Let x = The number sought. If there are seven times as many candles as nine, there must be $x = 7 \times 9 = 63$ candles.

764. (c) Let D = The unknown distance between farm-houses, in miles. Recall that, given a uniform rate, Distance = Rate × Time, or: D = RT. We know R = 12 mph and T = 42 minutes. Now convert minutes into hours by establishing. The ratio 42 minutes is to 60 minutes as x hours is to 1 hour or: $\frac{42}{60} = \frac{x}{1}$. Cross-multiplying: $42 = 60x$ or $x = 0.7$ hour. Thus D = $(12)(.7) = 8.4$ miles.

765. (d) Let T = The time, in minutes, it takes the snail to cross the road. Notice that the information about the car is irrelevant (although we hope the snail crosses safely). Since this is a uniform motion problem, we have: Distance = Rate × Time or D = RT. First, convert the Rate units to $\frac{feet}{minute}$ by establishing the ratio: 4 inches is to 12 inches as x is to 1 foot or $\frac{4}{12} = \frac{x}{12}$. So $x = 0.33$ feet and R = 0.33 $\frac{feet}{minute}$. Substituting: $19.8 = 0.33T$ or $T = \frac{19.8}{.33}$. Thus T = 60 minutes.

766. (a) This is the same as the equation provided; each score is divided by three.

767. (a) We are solving for x = The % saline concentration in the final 20 gallon mixture. The final amount equals the initial amount plus the amount added or F = I + A. We know the initial amount of 15% solution is I = 13 gallons and the amount of pure water added is A = 7 gallons. Noting that pure water is a 0% saline solution, we have: $x(20) = 0.15(13) + 0.00(7)$. Simplifying: $20x = 1.95$. Thus $x = 9.75\%$.

768. (c) We are trying to find T = Number of minutes it will take the tyre to completely deflate. The formula to use is Pressure = Rate × Time, or P = RT. In terms of T this becomes $T = \frac{P}{R}$. We are given P = 36 psi, R = 3 $\frac{psi}{minutes}$, therefore T = $\frac{36}{3}$. Thus T = 12 minutes.

SET 49

769. (d) Let x = The number of months needed to pay off the loan. First convert five-sixths of a year to x months by establishing the ratio: 5 is to 6 as x is to 12 or: $\frac{5}{6} = \frac{x}{12}$. Cross-multiplying: $(5)(12) = 6x$ or $x = 10$ months. Counting 10 months forward we arrive at the answer, the end of December.

770. (c) Let x = The amount of pure thinner to be added. The total number of gallons in the final mixture $F = (5 + x)$ gallons must equal the initial amount plus the amount added, or $F = I + A$. In terms of the % concentration of paint thinner this becomes: $0.09(5 + x) = 0.03(5) + 1.00x$. Simplifying: $0.45 + 0.09x = 0.15 + x$ which reduces to $0.30 = 0.91x$ or $x = 0.33$ gallon. So $F = (5 + 0.33) = 5.33$ gallons.

771. (c) Let x = The number sought. If there were 10 times as many pieces of chicken as 6, there must be $x = 6 \times 10 = 60$ pieces of chicken.

772. (b) We are trying to find x = The number of birds originally in the oak tree. Ten more birds landed means there are now $x + 10$ birds, a total of four times as many birds means the oak tree now has $4(x + 10)$ birds. In the maple tree: 16 less than 12 times as many bird as the oak tree had to begin with means there are $12x - 16$ birds in it. Setting the two equations equal we have: $4(x + 10) = 12x - 16$. Simplifying: $4x + 40 = 12x - 16$ or $8x = 56$. Thus $x = 7$ birds.

773. (b) We are looking for D = The number of dimes in the jar. Twice as many pennies as dimes means there are $P = 2D$ number of pennies. The total amount of coins in the jar is $T = 0.25Q + 0.10D + .05D + .01P$. Substituting: $4.58 = 0.25(13) + 0.10(D) + .05(5) + .01(2D)$. Simplifying: $4.58 = 3.25 + 0.10D + 0.25 + .02D$ which reduces to $0.12D = 1.08$. Thus $D = 9$ dimes.

774. (d) We want to know the value of P = The number, in feet, of the lot's perimeter, which is the sum of its sides or: $P = S1 + S2 + S3 + S4$. We are given $S1 = 130$, $S2 = 1.66(260)$, $S3 = 2(130)$, and $S4 = 1.66(260)$. Substituting: $P = 130 + 433 + 260 + 433$ which reduces to $P = 1,256$ feet.

775. (b) We are looking for x = The hiker's total trip time in hours, which will be twice the time it takes to get from his car to the lake. We are told that he travels 2 hours over smooth terrain. The time to walk the rocky trail is found by using the Distance formula (Distance = Rate × Time) and rearranging to solve for: $T = \frac{D}{R}$. Substituting, we get: $T = \frac{5}{2} = 2.5$ hours. Therefore $x = 2(2 + 2.5) = 9$ hours.

776. (d) Total pressure is equal to $P = O + N + A$. We are given: $N = 4$, $O = \frac{1}{2}N = 2$, and $A = \frac{1}{3} = \frac{2}{3}$; so $P = 2 + 4 + \frac{2}{3}$; $P = 6\frac{2}{3}$ psi.

1000 MATH PROBLEMS >>> *Answers*

777. (b) We are seeking D = Number of feet away from the microwave where the amount of radiation is $\frac{1}{16}$ the initial amount. We are given: radiation varies inversely as the square of the distance or: $R = 1 \div D^2$. When D = 1, R = 1, so we are looking for D when $R = \frac{1}{16}$. Substituting: $\frac{1}{16} = 1 \div D^2$. Cross-multiplying: $(1)(D^2) = (1)(16)$. Simplifying: $D^2 = 16$, or D = 4 feet.

778. (b) Let F = The Final amount of money Giri has now, which is the Initial amount less the Purchases amount or F = I − P. Since Giri has 19% of his money left he has spent 81% on purchases plus Rs. 1.29 for coffee. Therefore: P = 0.81(50) + 1.29 = 41.79. So F = 50 − 41.79 = Rs. 8.21 left.

779. (d) Let x = The amount which the five tip-paying coworkers must contribute. All seven coworkers must pay Rs. 48.72 ÷ 7 = Rs. 6.96 each. The 15% tip equals Rs. 48.72(.15) = Rs. 7.31. So the additional amount each of the five tip-paying coworkers must contribute is Rs. 7.31 ÷ 5 = Rs. 1.46. Thus x = 6.96 + 1.46 or x = Rs. 8.42.

780. (d) Let F = The Final amount of money Raj will receive, which is his Salary less the Placement fee or F = S − P. We are given S = Rs. 28,000 and P = $\frac{1}{12}$S. substituting F = 28,000 − $\frac{1}{12}$ or F = 28,000 − 2,333. Thus F = Rs. 25,667.

781. (d) The problem asks you to add $\frac{1}{3}$ of 60 and $\frac{2}{5}$ of 60. Let x = The number sought. We have: $x = \frac{1}{3}(60) + \frac{2}{5}(60)$ or $x = 20 + 24$. Thus $x = 44$.

782. (a) The 90% discount is over all three items; therefore the total price is $(a + b + c) \times 0.9$. The average is the total price divided by the number of computers: $0.9 \times \frac{(a+b+c)}{3}$.

783. (b) Let C = The number of cherries. It is given that 3 apples and 6 oranges equals $\frac{1}{2}$C or $9 = \frac{1}{2}$C. Therefore C = 2(9) = 18.

784. (b) Let L = The number of gallons of gas lost, which is equal to the Rate of loss times the Time over which it occurs or L = RT. Substituting: $L = (7)(\frac{1}{3}) = 2\frac{1}{3}$ gallons. Notice that the 14 gallon tank size is irrelevant information in this problem.

SET 50

785. (c) The problem can be restated as: 5 hours is to 24 hours as x% is to 100%. This is the same as: $\frac{5}{24} = \frac{x}{100}$.

1000 MATH PROBLEMS >>> Answers

786. (c) Let x = The number of kgs of white flour. The problem can be restated more usefully as: 5 parts is to 6 parts as x kgs is to 48 kgs or $\frac{5}{6} = \frac{x}{48}$. Cross-multiplying: $(5)(48) = 6x$ or $x = \frac{240}{6}$. Thus $x = 40$.

787. (a) Let x = The extra amount Hanif will earn by charging 25 paise instead of 10 paise per glass, which must be the difference in his total sales at each price. Therefore x = Sales at 0.25 per glass − Sales at 0.10 per glass. Substituting: $x = 0.25(7) − 0.10(20) = 1.75 − 2.00$. Thus $x = −.25$. Hanif will lose money if he raises his price to 25 paise per glass.

788. (b) Let J = The number of miles away from school that Dinesh lives. We are given $R = 5$ and $RA = \frac{1}{2}R = 2.5$. The distance between Raj and Ravi's house is $(5 − 2.5)$. Since Dinesh lives half way between them we have: $D = R + \frac{1}{2}(5 − 2.5)$. Substituting $D = 2.5 + \frac{1}{2}(2.5)$. Thus $D = 3.75$ miles.

789. (b) Let V = Veena's annual salary. We know: Praveena's earnings are six times Anjali's or $P = 6A$, Anjali earns five times more than Beena or $A = 5B$, and Beena earns Rs. 4,000 or $B = 4000$. Working backwards we have: $A = 5(4,000) = 20,000$, $P = 6(20,000) = 120,000$. Finally, we are told that Veena earns $\frac{1}{2}$ as much as Praveena or $V = \frac{1}{2}P$ so $V = \frac{1}{2}(120,000)$. Thus V = Rs. 60,000.

790. (b) Let G = Geeta's age. One-fourth Geeta's age taken away from Yogita's becomes $Y − \frac{1}{4}G$, twice Geeta's age becomes $2G$. Combining terms: $Y − \frac{1}{4}G = 2G$ which simplifies to $Y = (2 + \frac{1}{4})G$ or $Y = 2.25G$. Substituting: $9 = 2.25G$ or $G = 4$. Geeta is 4 years old.

791. (a) Let x = The number of kgs of chocolate to be mixed. We know the mixture's total cost is the cost of the chocolates plus the cost of the caramels or $M = A + B$. In terms of x: $M = 3.95(x + 3)$, $A = 5.95x$, while $B = 2.95(3)$. Combining terms: $3.95(x + 3) = 5.95x + 2.95(3)$. Simplifying: $3.95x + 11.85 = 5.95x + 8.85$ or $11.85 − 8.85 = (5.95 − 3.95)x$ which becomes $2x = 3$. Thus $x = 1.5$ kgs.

792. (b) We want to find S = The rate Sherly charges to mow a lawn in rupees per hour. We are given Kiran's rate, which is $K = 7.50$, and we are told that $S = 1.5 K$. Substituting: $S = 1.5(7.50)$. Thus S = Rs. 11.25 per hour.

793. (c) The problem is to find H = Years Hira will take to save Rs. 1000. We are told Leela saves three times faster than Hira, a ratio of 3:1. Therefore, $3L = H$. We are given $L = 1.5$ years. Substituting: $3(1.5) = H$ or $H = 4.5$ years.

794. (a) The unknown is W = FFinal number of gallons of water in the tank. The initial amount of water in the 50% solution is: $I = 0.50(5) = 2.5$ gallons. Half of the tank is drained: D = 0.50(.50)(5) = 1.25 gallons. Then 2 gallons of water are added: A = 2 gallons. This is expressed as W = I − D + A. Substituting: W = 2.5 − 1.25 + 2, or W = 3.25 gallons.

795. (c) The problem is to find x = Cost per kg of the total mixture. We know the total mixture cost must equal the sum of the individual costs or: T = S + M. We also know that T = (8 + 18)x, S = (3)(8), and M = (.50)(18). Combining terms: 26x = 24 + 9. Simplifying: 26x = 33. Thus x = 33 ÷ 26, or x = Rs. 1.27 per kg.

796. (a) Since the distance from the wall is known, the formula would be: $\frac{x}{5} + 2 = 10$. To find x, start by subtracting the 2, giving 8; then: $\frac{x}{5} = 8$, and 8 × 5 = 40; therefore, x = 40.

797. (b) The problem is to find J = The number of Jona's toys. It is given that J = W + 5, W = $\frac{1}{3}$T, T = 4E, and E = 6. Substituting: T = 4(6) = 24, W = $\frac{1}{3}$(24) = 8, and J = 8 + 5. Thus J = 13 toys.

798. (b) Let x = The number of hours to paint the sign if both worked together. In 1 hour Sushma can do $\frac{1}{6}$ of the job while Jona can do $\frac{1}{5}$ of the job. In terms of x this becomes: $\frac{1}{6}x + \frac{1}{5}x = 1$ (where 1 represents 100% of the job). Solving for x we have: $(\frac{1}{6} + \frac{1}{5})x = 1$ or $(\frac{5}{30} + \frac{6}{30})x = 1$. Simplifying $\frac{11}{30}x = 1$ or $x = \frac{30}{11}$. Thus x = 2.73 hours.

799. (a) Let x = The number of hours to decorate the window when both work together. In 1 hour Krishna can do $\frac{1}{3}$ of the job while Mona can do $\frac{1}{2}$ of the job. In terms of x we have: $\frac{1}{3}x + \frac{1}{2}x = 1$ (where 1 represents 100% of the job). Solving for x: $(\frac{1}{3} + \frac{1}{2})x = 1$ or $(\frac{2}{6} + \frac{3}{6})x = 1$. Simplifying: $\frac{5}{6}x = 1$ or $x = \frac{6}{5}$. Thus x = 1.2 hours.

800. (c) Let D = The distance travelled between the two cities. Using the formula Distance = Rate × Time, or D = RT, we have: D = 90(3.25) or D = 293 miles.

SET 51

801. (b) The problem asks us to find N = Nakul's age in years. We are given: R = 3N, Ram = 0.5N, and R + R + N = 117. Substituting: 3N + 0.5N + 1N = 117. Simplifying: (3 + 0.5 + 1)N = 117, which becomes 4.5N = 117. Thus N = 117 ÷ 4.5, or N = 26 years.

802. (d) Let I = The amount of interest earned. Substituting: I = (300)(7$\frac{3}{4}$%)($\frac{30}{12}$). Simplifying: I = (300)(.0775)(2.5). Thus I = Rs. 58.13.

803. (b) We must write the problem as an equation: J = 6K and J + 2 = 2(K + 2), so 6K + 2 = 2K + 4, which means K = $\frac{1}{2}$. J = 6K, or 3.

1000 MATH PROBLEMS >>> Answers

804. (c) M = 3N; 3N + N = 24, which implies that N = 6 and M = 3N, which is 18. If Nitin catches up to Mithilesh is typing speed, then both M and N will equal 18, and then the combined rate will be 18 + 18 = 36 pages per hour.

805. (b) Let x = The percent of students who are failing. The percentages must add up to 100. Therefore 13% + 15% + 20% + 16% + x = 100%, or x = 100 − 64. So x = 36%.

806. (d) We are seeking P = The initial Principal amount. Recall that Principal × Rate × Time = Interest or PRT = I. In one year P + I = 1,000, so I = 1,000 − P. Therefore PRT = 1,000 − P. Substituting: P(.05375)(1) = 1,000 − P. Rearranging: P + P(.05375) = 1,000. Simplifying: P(1 + .05375) = 1,000, or P = 1,000 ÷ 1.05375. Thus P = Rs. 949.

807. (a) If the mixture is $\frac{2}{3}$ raisins it must be $\frac{1}{3}$ nuts or: $4(\frac{1}{3})$ = 1.3 kgs.

808. (d) To solve: rearrange, convert units to feet, then plug in the values; A = (f times D) ÷ 1 = (0.5 × 3000) ÷ 0.25; A = 6000 feet.

809. (d) Let x = the number of hours it will take Pran to sew the dress. In 1 hour Chetna can do $\frac{1}{15}$ of the job, while Pran can do $\frac{1}{x}$ of the job. Since they can do the job together in 9 hours, we have: $\frac{1}{15}9 + \frac{1}{x}9 = 1$ (where 1 represents 100% of the job). Simplifying: $\frac{9}{15} + \frac{9}{x} = 1$ or $\frac{9}{x} = 1 - \frac{9}{15}$. Therefore $\frac{9}{x} = \frac{6}{15}$. Cross-multiplying: $6x = (9)(15)$ or $x = 22.5$ hours.

810. (b) Let x = The number of pages in last year's directory. 114 less than twice as many pages as last year's means $2x - 114$, so the equation becomes: $2x - 114 = 596$ or $2x = 710$. Thus $x = 355$ pages.

811. (b) The problem is to find M = Total number of miles travelled. We are given: D1 = 300 miles; D2 = $\frac{2}{3}$(D1); D3 = $\frac{3}{4}$(D1 + D2). We know that M = D1 + D2 + D3. Substituting: M = 300 + $\frac{2}{3}$(300) + $\frac{3}{4}$[300 + $\frac{2}{3}$(300)]. Simplifying: M = 300 + 200 + 375. Thus M = 875 miles.

812. (b) We are seeking F = The number of gallons of water in the barrel after the thunderstorm. This Final amount of water equals the Initial amount plus the Added amount, or F = I + A. We know: I = 4 gallons, and using the formula A = Rate × Time, we solve for A = ($\frac{6 \text{ gallons}}{\text{day}}$)($\frac{1}{3}$ day) = 2 gallons. Substituting: F = 4 + 2 = 6 gallons.

813. (a) We want to solve for J = Jenny's age on her next birthday. Jenny will be one-fourth Hari's age means J = 0.25H, Hari will be twice as old as Nishi means H = 2N, Nishi will be 10 years younger than Raj means N = R − 10.

1000 MATH PROBLEMS >>> Answers

We know Raj is 28 years old so R = 28. Now work backwards through the problem: N = 28 − 10 = 18, H = 2(18) = 36. Finally, J = 0.25(36) = 9 years.

814. (c) We know the three species of songbirds total 120 or A + B + C = 120. We know A = 3B and B = $\frac{1}{2}$C. This means A = $\frac{3}{2}$C. Substituting and solving for C we have: $\frac{3}{2}$C + $\frac{1}{2}$C + C = 120; 3C = 120; C = 40.

815. (a) Two equations must be used. 2B + 2H = 32; B = 7H. This gives 14H + 2H = 32; 16H = 32. H = 2, and B = 14. A = B × H; A = 2 × 14 = 28 m².

816. (d) Let x = The amount of pure cocoa to be added. We know the amount of cocoa in the 12% mix = (10)(0.12), and the total amount of cocoa in the final mix = (10 + x)(0.18). Combining terms: x + (10)(.12) = (10 + x)(0.18) which becomes x − 0.18x = 1.8 − 1.2. Thus x = 0.73 kgs.

SECTION 6 : GEOMETRY

SET 52

817. **(d)** A square is a special case of all of these figures except the trapezoid. A square is a parallelogram, because its opposite sides are parallel. A square is a rectangle because it is a quadrilateral with 90-degree angles. A square is a rhombus because it is a parallelogram with all sides equal in length. However, a square is not a trapezoid because a trapezoid has only two sides parallel.

818. **(b)** A cube has 4 sides, a top, and a bottom, which means that it has 6 faces.

819. **(a)** A polygon is a plane figure composed of 3 or more lines.

820. **(d)** An acute angle is less than 90 degrees.

821. **(a)** A straight angle is exactly 180 degrees.

822. **(c)** A right angle is exactly 90 degrees.

823. **(b)** Because parallel lines never intersect, choice **(a)** is incorrect. Perpendicular lines do intersect so choice **(c)** is incorrect. Choice **(d)** is incorrect because intersecting lines have only one point in common.

824. **(a)** A triangle with two congruent sides could either be isosceles or equilateral. However, because one angle is 40 degrees, it cannot be equilateral (the angle would be 60 degrees).

825. **(c)** The sum of the angles on a triangle is 180 degrees. The two angles given add to 90 degrees, showing that there must be a 90 degree angle. It is a right triangle.

826. **(a)** All of the angles are acute, and all are different. Therefore, the triangle is acute scalene.

827. **(c)** A trapezoid is the only one that does not have parallel lines by definition.

828. **(c)** The sum of the measure of the angles in a triangle is 180 degrees. 70 degrees + 30 degrees = 100 degrees. 180 degrees − 100 degrees = 80 degrees. Therefore, angle C is 80 degrees.

829. **(d)** A quadrilateral is a polygon with four sides. This eliminates the triangle. Rectangles, squares, and parallelograms all have two parallel sides, but only a parallelogram can have an angle that measures something other than 90 degrees.

830. **(c)** If the pentagons are similar, then the two different pentagons will have similar proportions. Because AB is similar to FG, and AB = 10, and FG = 30, the

second pentagon is 3 times as large. Therefore, IH is 3 times as large as CD, which gives 15.

831. (c) The greatest area from a quadrilateral will always be a square. Therefore, a side will be 24 ÷ 4 = 6 feet. The area is 6^2 = 36 square feet.

832. (b) The area of the square is 4 × 4 = 16 square feet; the area of the circle is $\pi(2^2) = \pi 4$. The difference is $16 - 4\pi$.

SET 53

833. (c) The perimeter is 4 × 4 for the square, and d for the circle. This is a difference of 16 − 4.

834. (c) The area is length × width, or in this case, A × 3 × A, $3A^2$.

835. (a) Use the Pythagorean theorem: $a^2 + b^2 = c^2$. The hypotenuse is found to be 5: $3^2 + 4^2 = 9 + 16 = 25$, and the square root of 25 is 5. The sum of the sides is the perimeter: 3 + 4 + 5 = 12.

836. (d) There are four sides of 4, and two sides of 8. Therefore, the perimeter is (4 × 4) + (2 × 8) = 32.

837. (c) The perimeter is the sum of the triangle's two legs plus the hypotenuse. Knowing two of the sides, we can find the third side, or hypotenuse (h), using the Pythagorean theorem: $a^2 + a^2 = h^2$, which simplifies to $2a^2 = h^2$. So $h = \sqrt{2a^2}$. This means the perimeter is $2a + \sqrt{2a^2}$.

838. (a) The first step in solving the problem is to subtract 86 from 148. The remainder, 62, is then divided by 2 get 31 feet.

839. (d) There are two sides 34 feet long and two sides 20 feet long. Using the formula P = 2L + 2W will solve this problem. Therefore, you should multiply 34 times 2 and 20 times 2, and then add the results: 68 + 40 = 108.

840. (d) The rectangular portion of the doorway has two long sides and a bottom: (2 × 10) + 4 = 24. The arc is $\frac{1}{2}\pi d = 2\pi$.

841. (c) The sum of the side lengths is 7 + 9 + 10 = 26.

842. (d) Find the slant height using the Pythagorean theorem: $6^2 + 8^2 = 36 + 64 = 100$. The square root of 100 is 10, so that is the measure of the missing side (the slant height). The perimeter is therefore (2 × 18) + (2 × 10) = 56.

843. (d) The curved portion of the shape is $\frac{1}{4}\pi d$, which is 4π. The linear portions are both the radius, so the solution is simply $4\pi + 16$.

844. (c) The perimeter is equal to 4 + 7 + 13 = 24.

845. (d) The sum of the measurements is the perimeter. This is 4 × 5 inches + 2 × 7 inches.

846. (c) The perimeter is equal to (2 × 4) + (2 × 9) = 26.

1000 MATH PROBLEMS >>> Answers

847. (d) The angle between a and d and the angle adjacent to it are complementary, so the adjacent angle is 60 degrees. The angle between c and the bottom line is also 60 degrees, so d and c must be parallel.

848. (c) The 60-degree angle and the angle between angle F and the 90-degree angle are vertical angles, so this angle must be 60 degrees. The 90-degree angle is supplementary to angle F and the adjacent 60 degree angle. 180 degrees − 90 degrees − 60 degrees = 30 degrees.

SET 54

849. (b) The angle between section a and wall 1 and the angle between section c and wall 2 are vertical angles. This means that the angle is the same as the section a-wall 1 angle, 45 degrees.

850. (d) Since the rows are parallel, the 55-degree angle is equal to the angle which is supplementary to both the 30-degree and the car's turning angle: 180 − 30 − 55 = 95 degrees.

851. (b) The shortest side is opposite the smallest angle. The smallest angle is 44 degrees, angle ABC. Therefore, the shortest side is AC.

852. (d) If two angles are 60 degrees, the third must also be 60 degrees. This is an equilateral triangle. All sides are therefore equal.

853. (b) Because the lines are parallel, angles are the same for both intersections. Opposite angles are congruent, making 1 and 7 congruent.

854. (c) The sum of the angles is 180 degrees. The two angles given add to 103 degrees, so angle ABC = 180 − 103 = 77 degrees.

855. (a) The distance between Plattville and Quincy is the hypotenuse of a right triangle with sides of length 80 and 60. The length of the hypotenuse equals $\sqrt{80^2 + 60^2}$, which equals $\sqrt{6400 + 3600}$, which equals $\sqrt{10,000}$, which equals 100 miles.

856. (a) The side opposite the largest angle is the longest side. In this case, it is side AB.

857. (b) If the angle's complement is half the angle's size, then it is $\frac{1}{3}$ of 90 degrees. That means that the angle is $\frac{2}{3}$ of 90 degrees, or 60 degrees.

858. (c) Complementary angles add to 90 degrees. Therefore, the complementary angle is 90 − 36 = 54 degrees.

859. (d) Supplementary angles add to 180 degrees. Therefore, the supplementary angle is 180 − 137 = 43 degrees.

860. (c) The equation is $3x = 2(90 − x)$. This reduces to $5x = 180$; $x = 36$ degrees.

861. (c) The equation is $180 − x = 3(90 − x)$. This reduces to $2x = 90$ degrees; $x = 45$ degrees.

862. (d) This is the only choice that includes a 90-degrees angle.

1000 MATH PROBLEMS >>> Answers

863. (c) 135 and its adjacent angle within the triangle are supplementary, so 180 − 135 = 45 degrees. Angle B and the remaining unknown angle inside the triangle are vertical, so the angle within the triangle's measure is needed: 180 − 60 − 45 = 75 degrees, so angle B is also 75 degrees.

864. (d) If the figure is a regular decagon, it can be divided into ten equal sections by lines passing through the centre. Two such lines form the indicated angle, which includes three of the ten sections. $\frac{3}{10}$ of 360 degrees is equal to 108 degrees.

SET 55

865. (c) PQ and RS are intersecting lines. The fact that angle POS is a 90-degree angle means that PQ and RS are perpendicular, indicating that all the angles formed by their intersection, including ROQ, measure 90 degrees.

866. (b) The dimensions of triangle MNO are double those of triangle RST. Line segment RT is 5 cm; therefore line segment MO is 10 cm.

867. (b) Angles 1 and 4 are the only ones NOT adjacent to each other.

868. (d) The perimeter is the total length of all sides. In a square, all four sides are of equal length, so the perimeter is 4 + 4 + 4 + 4, or 16.

869. (c) This can be divided into a rectangle and two half-circles. The area of the rectangle is 4(8) = 32 square feet. The diameter of the half-circles corresponds with the height and width of the rectangle. Therefore, the area of the circles is $\frac{2^2}{2}\pi = 2\pi$ and $\frac{4^2}{2} = 8\pi$. Therefore, the answer is 32 + 10π.

870. (c) An algebraic equation must be used to solve this problem. The shortest side can be denoted s. Therefore, $s + (s + 2) + (s + 4) = 24$. $3s + 6 = 24$, and $s = 6$.

871. (c) The area of a circle is A = (π)(r²); (π)(r²) = 16π; r = 4. The perimeter of a circle is P (2)(π)(r). P = (2)(π)(4) = 8π inches.

872. (b) The length of one side of a square is the perimeter divided by 4: 60 ÷ 4 = 15.

873. (a) The sum of the sides equals the perimeter: 3 sides × 3 inches + 2 sides × 5 inches = 19 inches.

874. (c) In order to find the perimeter, the hypotenuse of the triangle must be found. This comes from recognizing that the triangle is a 5 − 12 − 13 triangle, or by using the Pythagorean theorem. Therefore, 5 + 12 + 13 = 30.

875. (c) DE is 2.5 times greater than AB; therefore, EF is 7.5 and DF is 10. Add the three numbers together to arrive at the perimeter.

876. (b) Since the triangle is a right isosceles, the non-right angles are 45 degrees.

877. (a) To get the height of the triangle, use the Pythagorean theorem: $6^2 + height^2 = 10^2$. The height equals 8. Then 5 is plugged in for the base and 8 for the height in the area equation $A = \frac{bh}{2}$, which yields 20 square units.

1000 MATH PROBLEMS >>> Answers

878. (a) Because the curve opens downward, it must have a $-x^2$ term in it. Because the curve goes to the point (0, 4), the answer must be must be **(a)**.

879. (b) Since the 5-inch side and the 2.5-inch side are similar, the second triangle must be larger than the first. The two angles without congruent marks add up to 100 degrees, so 180 − 100 = 80 degrees. This is the largest angle, so the side opposite it must be largest, in this case side B.

880. (c) If angle 1 is 30°, angle 3 must be 60° by right triangle geometry. Because the two lines are parallel, angles 3 and 4 must be congruent. Therefore, to find angle 5, angle 4 must be subtracted from 180°. This is 120°.

SET 56

881. (c) The sum of the angles of this triangle must add up to 180 degrees. Also, the angles at the base of the triangle (call them b and c) are supplementary angles. This means angle b = 180 − 80 = 100 degrees, and angle c = 180 − 150 = 30 degrees. Thus, angle a must equal 180 − 130 = 50 degrees.

882. (c) In a parallelogram, adjacent angles are supplementary. Therefore, the ABC is equal to 180 − 88 = 92 degrees.

883. (d) Because one base angle of the triangle is 70 degrees, the other must be 70 degrees also. Therefore, the vertex angle is 180 − 140 = 40 degrees.

884. (d) The diameter of the fan blade will be 12 inches. The circumference is πd, giving a circumference of 12π inches.

885. (c) The equation is $2\pi r = \frac{1}{2}\pi r^2$. One r term and both π terms cancel, leaving r = 4.

886. (b) The circumference of a circle is πd. The diameter is 5 inches, giving a circumference of 5π inches.

887. (a) The top rectangle is 3 units wide and 3 long, for an area of 9. The bottom rectangle is 10 long and 23 wide, for an area of 230. These add upto be 239.

888. (c) The perimeter is the total of the length of all sides. Since the figure is a rectangle, opposite sides are equal. This means the perimeter is 5 + 5 + 2 + 2 = 14.

889. (c) 5 × 3 × 8 is 120. 120 ÷ 3 = 40.

890. (a) When the 2-by-2 squares are cut out, the length of the box is 3, and the width is 6. The height is 2. The volume is 3 × 6 × 2, or 36.

891. (c) The formula for area is Area = Length × Width, in this case, 64.125 = 9.5 × Width, or 6.75.

892. (c) Each 9-foot wall has an area of 9 × 8 or 72 square feet. There are two such walls, so those two walls combined have an area of 72 × 2 or 144 square feet. Each 11-foot wall has an area of 11 × 8 or 88 square feet, and again there are two such walls: 88 times 2 equals 176. Finally, add 144 and 176 to get 320 square feet.

1000 MATH PROBLEMS >>> Answers

893. (d) The area is width times length, in this case, 5 times 7, or 35 square feet.

894. (a) If the side of the barn is 100 feet, then the opposite side of the rectangle is made of 100 feet of fence. This leaves 100 feet of fence for each of the other 2 sides. The result is a square of area $100^2 = 10,000$ square feet.

895. (b) The surface area of the walls in four walls of 120 square feet. This gives 480 square feet. The area of the door and window to be subtracted is $12 + 21$ square feet = 33 square feet. Therefore, 447 square feet are needed. This would be 5 rolls.

896. (c) The area is $\frac{1}{2}$ base × height. This gives $\frac{1}{2}(4)(8) = 16$.

SET 57

897. (a) The area of the parallelogram can be found one of two ways. The first would be using a formula, which is not provided. The second is by splitting the parallelogram into two triangles and a rectangle. The rectangle would have an area of 36 square feet. The area of the triangles is 3(4). This gives a total area of 48.

898. (b) Area is equal to base times height. $2 \times 4 = 8$.

899. (d) The area is the width times the length—in this case, 10×8, or 80 square feet.

900. (b) This must be solved with an algebraic equation. $L = 2W + 4$; $3L + 2W = 28$. Therefore, $6W + 12 + 2W = 28$; $8W = 16$; $W = 2$. $L = 8$. $2 \times 8 = 16$ square inches.

901. (c) The longest object would fit on a diagonal from an upper corner to a lower corner. It would be in the same plane as one diagonal of the square base. This diagonal forms one side of a right triangle whose hypotenuse is the length of the longest object that can fit in the box. First, use the Pythagorean theorem to find the length of the diagonal of the square base; $3^2 + 3^2 = c^2$; $c = \sqrt{18}$. Next, find the length of the second side, which is the height of the box. Since the square base has an area of 9, each side must be 3 feet long; we know the volume of the box is 36 cubic feet so the height of the box (h) must be $3 \times 3 \times h = 36$, of $h = 4$ feet. Finally, use the Pythagorean theorem again to find the hypotenuse, or length of the longest object that can fit in the box; $\sqrt{18}^2 + 4^2 = h^2$; $18 + 16 = h^2$; $h = \sqrt{34} = 5.8$ feet.

902. (b) The lot will measure 200 feet by 500 feet, or 100,000 square feet in all (200 feet × 500 feet). An acre contains 43,560 square feet, so the lot contains approximately 2.3 acres (100,000 ÷ 43,560). At Rs. 9,000 per acre, the total cost is Rs. 20,700 (2.3 × Rs. 9,000).

903. (b) If the circle is 100π square inches, its radius must be 10 inches, using the formula $A = \pi r^2$. Side AB is twice the radius, so it is 20 inches.

1000 MATH PROBLEMS >>> Answers

904. (d) Each quilt square is $\frac{1}{4}$ of a square foot; 6 inches is $\frac{1}{2}$ a foot, $0.5 \times 0.5 = 0.25$ of a square foot. Therefore, each square foot of the quilt requires 4 quilt squares. 30 square feet × 4 = 120 quilt squares.

905. (c) There is a right triangle of hypotenuse 10 and a leg of 6. Using the Pythagorean theorem, this makes the height of the rectangle 8. The diagram shows that the height and width of the rectangle are equal, so it is a square. 8^2 is 64.

906. (a) The Pythagorean Theorem states that the square of the length of the hypotenuse of a right triangle is equal to the sum of the squares of the other two sides, so we know that the following equation applies: $1^2 + x^2 =$, so $1 + x^2 = 10$, so $x^2 = 10 - 1 = 9$, so $x = 3$.

907. (b) The triangles inside the rectangle must be right triangles. To get the width of the rectangle, use the Pythagorean theorem: $3^2 + x^2 = 18$. $x = 3$, and the width is twice that, or 6. This means the area of the shaded figure is the area of the rectangle minus the area of all 4 triangles. This is $12 \times 6 - 4 \times (\frac{1}{2} \times 3 \times 3)$. This equals 54.

908. (b) Since the canvas is 3 inches longer than the frame on each side, the canvas is 31 inches long (25 + 6 = 31) and 24 inches wide (18 + 6 = 24). Therefore the area is 31 × 24 inches, or 744 square inches.

909. (a) The formula for determining the area of a circle is πr^2, in this case, $\pi 3^2$ or 28.26 or approximately 28.

910. (b) The formula for finding the area of a triangle is: area = $\frac{1}{2}$(base × height), In this case, area = $\frac{1}{2}$(7.5 × 5.5) or 20.6 square inches.

911. (b) Each square foot of floor requires 4 tiles; the area of one tile is 36 square inches (6 × 6 = 36); the area of a square foot is 144 square inches (12 × 12 = 144). The square footage of the room is 180 square feet (12 ×15 = 180). Since each square foot requires 4 tiles, 180 × 4 = 720.

912. (d) The formula for the area of a circle is πr^2. In this case, $113 = 3.14 \times r^2$; r = 6; 6 × 2 = 12.

SET 58

913. (b) The height times the length gives the area of each side wall. Each side wall is 560 square feet; multiply this times 2 (there are two sides). The front and back each have an area of 448 square feet; multiply this times two. Add the two results to get the total square feet. 20 × 28 = 560 × 2 = 1120. 20 × 16 = 320 × 2 = 640. 640 + 1120 = 1760. The total square footage is 1760. Each gallon covers 440 square feet. 1760 ÷ 440 = 4.

914. (c) The area is width times length, in this case, 6 × 6 = 36 square feet.

1000 MATH PROBLEMS >>> Answers

915. (b) The area is the width times length, in this case, 6.5 × 8 or 52 square feet.

916. (d) The total area of the kitchen (11.5 × 11.5) is 132.25 square feet; the area of the circular inset ($\pi 3^2$) is 28.26 square feet: 132.25 − 28.26 = 103.99, or 104 square feet.

917. (a) The formula for area is: area = $\frac{1}{2}$(base × height) or $\frac{1}{2}$(75 × 167), making the area of the lot 6262.5 square feet.

918. (a) The area of each poster is 864 square inches (24 inches × 36 inches). Kavita may use four posters, for a total of 3456 square inches (864 × 4). Each picture has an area of 24 square inches (4 × 6); the total area of the posters should be divided by the area of each picture, or 3456 ÷ 24 = 144.

919. (d) The area of the picture is 475 square inches (25 × 19) and the area of the frame is 660 square inches. The difference—the area the mat needs to cover—is 185 square inches.

920. (d) The formula for finding the area of a triangle is A = $\frac{1}{2}$(base × height). In this case, A = $\frac{1}{2}$(20 × 30) or 300 square feet.

921. (d) The formula for finding the length of fabric is area = width × length. 54 inches is equal to $1\frac{1}{2}$ yards, so in this case the equation is 45 = 1.5 × length, or 45 ÷ 1.5 = length or 30 yards.

922. (a) The total area of the long walls is 240 square feet (8 × 15 × 2) and the total area of the short walls is 192 square feet (8 × 12 × 2). The long walls will require 2 cans of Moss and the short walls will require 1 can of Daffodil.

923. (d) The area of the house (30 feet × 50 feet) is 1,500 square feet. 1,500 × 2.53 = 3,795 or Rs. 3,795.

924. (c) First convert a tile's inches into feet. Each tile is 0.5 foot. The hallway is 10 feet wide, but since the lockers decrease this by 2 feet, the width of the tiled hallway is 8 feet. To get the area, multiply 72 by 8, which is 576 feet. Then divide that by the tile's area, which is 0.5 foot times 0.5 foot, or 0.25 foot. 576 ÷ 0.25 = 2,304 tiles.

925. (d) The equation for area of a rectangle is length times width, which in this case is 50 times 100, or 5,000 feet.

926. (c) Since the waiter folded the napkin in half to get a 5-by-5 inch square, the unfolded napkin must be 10 by 5 inches. Area is length times width, which is in this case 50 square inches.

927. (d) The area of the dark yard is the area of her yard minus the circle of light around the lamp. That is $20^2 − \pi \times 10^2$, or $400 − 100\pi$.

928. (d) Because the radius of the hemisphere is 3, and it is the same as half the base of the triangle, the base must be 6. Therefore, the area of the triangle is $\frac{1}{2}bh$.

1000 MATH PROBLEMS >>> Answers

= 12. The area of the circle is π^2 which is equal to 9π. Therefore, the half-circle's area is $\frac{9\pi}{2}$. Adding gives $\frac{9\pi}{2} + 12$.

SET 59

929. (b) The total area of the stores is $20 \times 20 \times 35 = 14,000$ square feet. The hallway's dimensions are 2,000 feet. The total square footage is 16,000.

930. (b) The surface area in square inches is: $2 \times [2 \times (16 \times 36) + 2 \times (24 \times 36) + 2 \times (16 \times 24)] - 9$. This reduces to $2 \times [1,152 + 1728 + 768] - 9$, or 7,287 square inches.

931. (b) The equation of the area of a circle is π times r^2. In this problem, r is 10, so the answer is 100π.

932. (b) The equation for the perimeter of a circle is πd. The perimeter should be divided by π to get the diameter, which is 18 feet.

933. (d) To get the surface area of the walls, the equation is circumference times height. To get the circumference, plug 10 feet into the equation $C = 2\pi r$. This gets a circumference of 20π. When this and the height of 8 feet are plugged into the surface area equation, the answer is 160π square feet.

934. (d) Since the two sides have different measurements, one is the length and one the width. The area of a rectangle is found by multiplying length times width: Length = 15 inches. Width = $15 \div 3 = 5$ inches. 5 inches $\times$ 15 inches = 75 square inches.

935. (a) The area of a triangle is $A = \frac{1}{2} (b \times h)$. Since $b = 2h$, we have $16 = \frac{1}{2} (2h)(h)$ or $h^2 = 16$; $h = 4$ inches.

936. (b) Area = Length $\times$ Width or $L \times 12 = 132$. Thus $L = 11$.

937. (a) Area is $\frac{1}{2}(b \times h)$. To get the height of the triangle, Pythagorean theorem is used: $3^2 + height^2 = 5^2$, so height = 4. When plugged into the area equation, an area of 6 square units is obtained for half of the triangle. Double this, and the answer is 12 square units.

938. (c) The Pythagorean theorem gives the height of the parallelogram as 4; the area is $8 \times 4 = 32$.

939. (b) Since the top and bottom are parallel, this figure is a trapezoid whose area = (Top + Bottom) (Height) $\div$ 2. Plugging in the values given we have: $(3 + 6)(10) \div 2 = 45$ square units.

940. (c) The Pythagorean theorem is used to find the length of the diagonal. $5^2 + 12^2 = 169$. The square root of 169 is 13.

941. (d) The shaded area is the difference between the area of the square and the circle. Because the radius is 1, a side of the square is 2. The area of the square is 2×2, and the area of the circle is $\pi(1^2)$. Therefore, the answer is $4 - \pi$.

1000 MATH PROBLEMS >>> Answers

942. (b) The area of the rectangle is 5(4); the area of the triangle is $\frac{1}{2}(3)(4)$. The sum is $20 + 6 = 26$.

943. (a) There is no difference between the square and parallelogram in area. The formula for both is base times height. Both have areas of 16 square inches.

944. (b) The formula for finding the area of a circle is $A = \pi r^2$. First, square the radius: 13 times 13 equals 169. Then multiply by the approximate value of π, 3.14, to get 530.66.

SET 60

945. (c) A circle is 360 degrees, so 40 degrees is one-ninth of a circle. Multiply the perimeter of the track, 360 feet, by one-ninth, to get the answer of 40 feet.

946. (a) The formula for finding circumference is πd, or $C = 3.14 \times 2.5$. The circumference is 7.85 inches.

947. (d) The longest path will be a circular path around the post. This will be equal to the circumference of the circle, which is $2\pi r$. This gives 40π feet.

948. (c) The formula for determining the circumference is $C = \pi d$, in this case, $C = 3.14 \times 24$ or 75.36 inches.

949. (d) The formula for circumference is $2\pi r$. In this case, the equation is $2 \times 3.14 \times 4.5 = 28.26$ or $28\frac{1}{4}$ inches.

950. (b) The formula for finding the diameter—the minimum length a spike would need to be—is $C = \pi d$, in this case, $43.96 = 3.14 \times d$. $43.96 \div 3.14 = 14$.

951. (b) The formula for determining the circumference of a circle is $C = 2\pi r$, in this case, $C = 2 \times 3.14 \times 25$ or 157.

952. (a) The formula for finding the diameter is circumference = the diameter $\times \pi$ ($C = \pi d$) or $18.84 = \pi d$ or $18.84 = \pi 6$. The diameter is 6.

953. (c) The formula for determining the area of a circle is $A = \pi r^2$. The area is 78.5 square inches. Therefore, the formula is $78.5 = \pi r^2$, and the radius (r) equals 5.

954. (a) The area of the dough is 216 square inches (18 × 12). To find the area of the cookie cutter, first find the radius. The formula is $C = 2\pi r$, in this instance, $9.42 = 2 \times 3.14 \times r$, or $9.42 = 6.28 \times r$. Divide 9.42 by 6.28 to find r, which is 1.5. The formula for area of a circle is (πr^2), in this case, 3.14×1.5^2, or 7.07 square inches. So, the area of the cookie cutter circle is 7.07 square inches. Divide 216 (the area of the dough) by 7.07 (the area of the cookie cutter) and the result is 30.55 or approximately 31 cookies.

955. (b) If the wheel has a diameter of 27 inches, the circumference is $\pi \times 27$. Multiplied by 100 turns, the distance Skyler biked is $2{,}700\pi$ inches.

956. (a) The perimeter is the distance around a polygon and is determined by the lengths of the sides. The total of the three sides of the lot is 320 feet.

1000 MATH PROBLEMS >>> Answers

957. (a) The perimeter of a polygon is the total of the lengths of its sides. Each of six sides are 12 inches long, or 6 × 12 = 72.

958. (b) The perimeter of the room is 36 feet (9 × 4); 36 ÷ 15 (the length of each garland) = 2.4. So Naveena will need 3 garlands.

959. (c) The perimeter of a square is determined by adding the lengths of all four sides; or, since the sides are the same length, multiplying one side by 4. Therefore, the circumference divided by 4 is the length of one side. So, 52 ÷ 4 = 13.

960. (a) The unpainted section of the cloth is $6\frac{2}{3}$ feet by $8\frac{2}{3}$ feet, because each side is shortened by 16 inches (an 8-inch border at each end of the side). To find the perimeter, add the length of all four sides, or 6.66 + 6.66 + 8.66 + 8.66 = 30.66 or $30\frac{2}{3}$.

SET 60

961. (c) To find the perimeter of a rectangle, add the lengths of all four sides. In this case, the sum is 102 feet. Divide by 3 to determine the number of yards: 102 ÷ 3 = 34.

962. (b) The height of the triangle is, from Pythagorean theorem, $10^2 + height^2 = 20^2$, which is the square root of 300. Plug that into the area equation, area = $\frac{(base \times height)}{2}$ and 20 for the base. This gives an answer of 10 times the square root of 300 cubic feet.

963. (b) Facing in the exact opposite direction is turning through an angle of 180 degrees. Therefore, the new compass reading will be 32 degrees south of west. This is equivalent to the complementary angle (90 − 32) = 58 degrees west of south.

964. (d) Since the ceiling and the floor are parallel, the acute angles where the stairs meet the ceiling and floor are equal. These are also supplementary to the obtuse angles, so 180 minus 20 = 160 degrees.

965. (d) A 90-degree angle is called a right angle.

966. (b) When the sum of two angles is 180 degrees, the angles are supplementary or supplemental to one another. To find the supplement subtract 35 degrees from 180 degrees, or 180 − 35 = 145.

967. (a) Since the bumpers are parallel, then the angle whose measure is needed must be equal to the angle in between the two equal angles, or 36 degrees.

968. (b) The ground creates a straight angle, which is 180 degrees. The obtuse angle is 180 − 54 = 126.

969. (d) A parallelogram is a quadrilateral in which parallel sides are of the same length (two sides are 20 feet, two are 17 feet) in which the opposite angles are not right angles.

1000 MATH PROBLEMS >>> Answers

970. (c) The vertex is the point of an angle.

971. (c) The triangle created by the chicken, the cow, and the bowl of corn is a right triangle; the 90 degree angle is at the point where the chicken is standing. Use the Pythagorean Theorem ($a^2 + b^2 = c^2$) to find the missing side of this right triangle. In this case, a = 60 × 60; c = 80 × 80 (the hypotenuse is c) or 3600 + b^2 = 6400. To solve, 6400 − 3600 = 2800; the square root of 2800 is 52.915, rounded to 53. The chicken will walk about 53 feet to cross the road.

972. (c) Since the ships are going west and north, their paths make a 90 degree angle. This makes a right triangle where the legs are the distances the ships travel, and the distance between them is the hypotenuse. Using the Pythagorean theorem, $400^2 + 300^2 = distance^2$. The distance = 500 miles.

973. (a) A parallelogram contains opposite angles that are the same size, in this case, A and C are each 70 degrees and B and D are each 110 degrees. If, however, the angles were each 90 degrees, the shape would be a square or a rectangle.

974. (d) The total number of degrees around the centre is 360. This is divided into 6 equal angles, so each angle is determined by the equation: 360 ÷ 6 or 60.

975. (c) An angle that is more than 90 degrees is an obtuse angle.

976. (c) The Pythagorean theorem is used to solve this problem. Forty feet is the hypotenuse, and 20 feet is the height of the triangle. $40^2 = 20^2 + w^2$; $w^2 = 1,200$; w = 34.6. This is closest to 35 feet.

SET 62

977. (a) Leela is walking in a right triangle, therefore use the Pythagorean theorem ($a^2 + b^2 = c^2$) to calculate the missing side: $9^2 + 9^2 = c^2$. So c = $\sqrt{162}$ or 12.7 feet.

978. (a) The angle that the ladder makes with the house is 75 degrees, and the angle where the house meets the ground must be 90 degrees, since the ground is level. Since there are 180 degrees in a triangle, the answer is 180 − 90 − 75 = 15 degrees.

979. (b) To bisect something, the bisecting line must cross halfway through the other line. 36 ÷ 2 = 18 feet.

980. (d) First find the number of yards it takes Mani to go one way: 22 × 90 = 1,980. Then double that to get the final answer: 3,960.

981. (c) First find the area of the brick wall: A = lw, or 10 ft × 16 ft = 160 sq. ft. Now convert 160 square feet to square inches—but be careful! There are 12 linear inches in a linear foot, but there are 144 (12^2) square inches in a square foot. 160 × 144 = 23,040 square inches. Divide this area by the area of one brick (3 inches × 5 inches = 15 sq in): $\frac{23,040}{15}$ = 1,536.

1000 MATH PROBLEMS >>> Answers

982. (b) The midpoint is the centre of a line. In this case, it is 400 ÷ 2, or 200 feet.

983. (d) When two segments of a line are congruent, they are of the same length. Therefore, the last block is 90 feet long (the same as the second block). To arrive at the total distance, add all the segments together; 97 + 2(90) + 3(110) + 90 = 697.

984. (a) A transversal line crosses two parallel lines. Therefore, in North Boulevard transverses Main Street, it also transverses Broadway.

985. (c) A perpendicular line crosses another line to form four right angles. This will result in the most even pieces.

986. (b) Consider the houses to be points on the straight line of the border. Points that lie on the same line are collinear.

987. (c) The area of a triangle is $(\frac{1}{2})bh$ or $A = \frac{1}{2}(4 \times 5)$ or 10.

988. (b) The perimeter is the distance around a polygon and is determined by the lengths of the sides. The total of the four sides of your yard is 360 feet. $70 \times 2 + 110 \times 2 = 360$. The gate subtracts 3 feet from the length. $360 - 3 = 357$.

989. (d) An obtuse angle is more than 90 degrees and less than 180 degrees.

990. (b) The volume of the water is $10 \times 10 \times 15 = 1500$ cubic inches. Subtracting 60 gets the answer, 1,440 cubic inches.

991. (c) The volume will equal the length times the width times the height (or depth) of a container: (12 inches) (5 inches) (10 inches) = 600 cubic inches.

992. (c) We are looking for the height (h) of water in the cylinder. First, find the volume of water in the hemisphere; $\frac{1}{2}[\frac{4}{3}(\pi)(r^3)] = 18\pi$, or about 56.5 cubic feet. The cylinder's water volume is the total water volume minus the volume in the hemisphere, or $170 - 56.5 = 113.5$ cubic feet. Next, solve for (h) using the formula for volume of a cylinder; $(\pi)(r^2)(h) = 113.5$; $(\pi)(3^2)(h) = 113.5$; $h = \frac{113.5}{(9)(\pi)}$. Thus h = 4 cubic feet.

SET 63

993. (d) The volume of concrete is 27 cubic feet. The volume is length times width times the depth (or height), or (L)(W)(D), so (L)(W)(D) equals 27. We're told that the length L is 6 times the width W, so L equals 6W. We're also told that the depth is 6 inches, or 0.5 feet. Substituting what we know about the length and depth into the original equation and solving for W, we get (L)(W)(D) = (6W)(W)(0.5) = 27. $3W^2$ equals 27. W^2 equals 9, so W equals 3. To get the length, we remember that L equals 6W, so L equals (6)(3), or 18 feet.

994. (a) The amount of water held in each container must be found. The rectangular box starts with 16 square inches × 9 inches = 144 cubic inches of water. The

1000 MATH PROBLEMS >>> Answers

cylindrical container can hold 44π cubic inches of water, which is approximately 138 cubic inches. Therefore, the container will overflow.

995. (b) The volume of the box is $144 + 32 = 176$ cubic inches. That divided by the base of the box gets the height, 11 inches.

996. (b) The volume of the briefcase is $24 \times 18 \times 6$ inches, or 2,596 cubic inches. The volume of each notebook is $9 \times 8 \times 1$ inches, or 72 inches. Dividing the volume of the briefcase by the volume of a notebook gets an answer of 36 notebooks.

997. (c) Think of the wire as a cylinder whose volume is $(\pi)(r^2)(h)$. To find the length of wire solve for h, in inches. One cubic foot $= (12)^3$ cubic inches $= 1,728$ cubic inches. Therefore $(\pi)(0.5^2)(h) = 1,728$; $h = \frac{4(1,728)}{\pi}$; $h = 6,912 \div \pi$.

998. (c) The proportion of Danny's height to his shadow and the pole to its shadow is equal. Danny's height is twice his shadow, so the pole's height is also twice its shadow, or 20 feet.

999. (d) The curve shown is $x^2 - 2x$.

1000. (c) Points B, C, and D are the only points in the same line and are thus also in the same plane.

NOTES

NOTES